Geology for Geotechnical Engineers

Geology for Geotechnical Engineers

J.C.HARVEY

Lecturer in Geotechnical Engineering, Plymouth Polytechnic, England

CAMBRIDGE UNIVERSITY PRESS

Cambridge

London New York New Rochelle

Melbourne Sydney

CAMBRIDGE UNIVERSITY PRESS
Cambridge, New York, Melbourne, Madrid, Cape Town, Singapore,
São Paulo, Delhi, Dubai, Tokyo, Mexico City

Cambridge University Press
The Edinburgh Building, Cambridge CB2 8RU, UK

Published in the United States of America by Cambridge University Press, New York

www.cambridge.org
Information on this title: www.cambridge.org/9780521288620

First published 1982
Re-issued 2011

A catalogue record for this publication is available from the British Library

Library of Congress Catalogue Card Number: 82-1289

ISBN 978-0-521-24629-3 Hardback
ISBN 978-0-521-28862-0 Paperback

Contents

Preface   vii

1 The Earth, its structure, and the forces acting inside it and
    on its surface   1
    The interior of the Earth and dynamic forces   1
    Weathering and decay of the Earth's surface   5
    Earth history   6

2 Rock-forming minerals and processes of rock formation and
    decomposition   10
    Rock-forming minerals   10
    Rock-forming processes   15
    Specific gravity of rocks   16
    Weathering and erosion   17

3 Rock types   18
    Igneous rocks   19
    Sedimentary rocks   31
    Metamorphic rocks   42
    Superficial deposits   46
    Conclusion   58

4 Geological structures, rock instability and slope movement   60
    Dip   60
    Folds   65
    Faults   68
    Joints   70
    Unconformity   70
    Igneous rock structures   72
    Landforms   73
    Rock instability and slope movement   77

5 Geological and geotechnical maps   90
    Geological maps   90
    Geotechnical maps   113

Contents

6 Engineering description of rocks   118
  Rock material description   118
  Rock material indices   119
  Rock mass description   126
  Rock mass indices   126

  Bibliography   131

  Index   133

Preface

This book describes the fundamental principles of geology that are needed for a study of geotechnics, the science of the physical properties of the material found in the ground in which civil engineers build their constructions. A brief study of the ground will show that it consists of hard or soft material, sometimes separate, sometimes both together. Geotechnical engineers define the hard material as rock, the soft as soil. The studies of the behaviour of these materials are called rock mechanics and soil mechanics, respectively. Geotechnics is the practical application of these sciences to forecast the behaviour of the ground when it is built upon or when excavations and tunnels are made into it.

To understand the behaviour of the ground the geotechnical engineer needs to know some essential geological facts. Rocks, and the soils which are derived from them, are described with special attention paid to those properties which are closely connected with their behaviour under mechanical stress and the chemical forces on the surface of the Earth which bring about decay of solid rock and convert it into soil. For a scientific study of the origins and classification of rock, the student has to learn very many names of the different chemical compositions of the materials which form rock, the names of very many rock types and their internal structures, and the origins and distributions of rocks throughout the world. Whilst a study of the details of the history of the Earth and its component materials can be of very great interest, civil engineers have in the past thought of geology as a science which has such a complicated terminology that they have tended to avoid it. They preferred to concentrate on soil, and their work in this direction over the last 50 years has produced a new branch of science – soil mechanics. Rock was the hard material found in some excavations in the ground. Sometimes it was welcome, making a solid base for the foundations of a building; sometimes it was a nuisance, causing rock trouble, difficult and expensive to excavate, or having unpredictable behaviour when exposed in excavations. Before 1950 civil engineers in general hoped that rock would not cause them any trouble. When there was rock trouble, civil engineers dealt with it as best they could and sometimes called in consultant geologists to help them. The advice they received was often unsatisfactory because the geologists were trained primarily to understand the Earth's history and processes, and to describe and classify rocks and the wide range of chemical substances they contained, including fossils and useful minerals. Geologists were not really interested in predicting rock behaviour, nor in a study of rotted and disintegrated rock making up soil. During the period 1950–1970 however, many geologists turned their attention to the mechanical properties of rock as the demand for reliable prediction of rock behaviour increased, as constructions became larger and the occurrence of rock trouble became more expensive and even led to the bankruptcy of some construction companies which were unable to meet target dates, or whose costs exceeded those stated in the contract. This application of the science to the practical problems of the construction industry became known as engineering geology. This new branch of geology was mostly concerned with rocks and civil engineers continued to be responsible for forecasting the behaviour of soils. The two cannot really be separated because they often occur close together and need to be tested as a whole instead of by separate groups of people. As geologists became more interested in engineering geology and civil engineers in rocks, the new science of geotechnics was born.

Because this treatment of the subject is mainly concerned with the mechanical behaviour of rock rather than its origins and variable chemical composition, the number of geological words has been carefully reduced to those essential for a sufficient knowledge of geological principles to understand the basis of geotechnics. Further information on geology can be found in the literature quoted in the Bibliography.

I should like to record my thanks to my colleague John Clatworthy of the Department of Civil Engineering, Plymouth Polytechnic, for advice and encouragement during the writing of this book, and for his collaboration in the geology training of civil engineering students during the past 20 years.

Plymouth   John Harvey
October, 1981

1 The Earth, its structure, and the forces acting inside it and on its surface

The interior of the Earth and dynamic forces

The Earth is not a rigid and static body, but is in a continual state of change, both inside and on the surface. Forces are acting to create new rock material and on the surface other forces are destroying the rock which has been formed in the past. The product of these destructive processes is known as soil, itself a new form of the material, so the forces of destruction may also be thought of as constructive forces. The word soil in geotechnics means any unconsolidated material in the ground and is not used in the same sense as that used by pedologists who study soil as a life-supporting material. The age of the Earth is at present believed to be at least 4500 million years. The ages of some rocks found on the surface of the Earth have been determined to be of the order of 3500 million years by using methods based on the radioactive decay of natural isotopes found in minerals which make up the body of the rock.

The interior of the Earth is believed to be built of concentric shells of rock material, named the crust and the mantle, which surround a central core (Fig. 1). Movement of material within the Earth is believed to be the cause of some of the surface processes which we experience. This movement may be the last phases of the turbulence associated with the creation of the planets in the solar system.

The average specific gravity of the Earth as determined by the mathematics of planetary motion is 5.5, but the specific gravity of the rock found on the surface is only about half this value, 2.7. Therefore the interior must be more dense. The evidence for the existence of a structure consisting of distinct shells of different specific gravities as shown in Fig. 1 comes from measurements on the rates at which shock waves from earthquakes travel through the Earth. There are different types of waves and they move in complicated paths through the interior. Results obtained from the study of these waves (seismology) have shown that there are relatively rapid changes in specific gravity at depths of 35, 700, and 2900 km. These changes are called discontinuities by geophysicists; this word is also used in geotechnics, but in a different sense, to indicate fractures and open spaces within rocks. These relatively rapid changes in specific gravity cause earthquake waves to be reflected and refracted; the subsequent paths followed by the waves can be detected and their study forms the basis of our knowledge of the interior of the Earth. Geotechnical exploration methods can determine the specific gravities of the rock below the surface as part of a site investigation programme to predict the behaviour of the rock below a construction site. Shock waves are generated by falling weights or small explosive charges. The same method on a larger scale is used in the exploration for oilfields.

The last 50 years of study of the continents, their structure and histories, have led geophysicists to believe that the continents and oceans are not permanently fixed in position but are moving extremely slowly across the surface of the Earth, a process known as continental drift. The ocean beds are not permanent. The evidence for this belief is that studies of the ages of rocks dredged from ocean beds show that the youngest rocks tend to be found in a world-encircling mid-ocean ridge system in the North and South Atlantic Oceans, Pacific Ocean, and between the Indian Ocean and Antarctica. The rocks farther away from the ridges are older than those within the ridges, from which it is inferred that rock is rising and being pushed sideways as more comes to the surface from the mantle. To maintain balance in the shape of the Earth there must be regions where rock is

Fig. 1. Interior of the Earth. The crust is drawn with exaggerated thickness to show the thickening below continents. Specific gravity values are given.

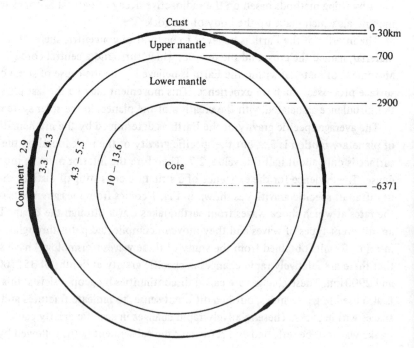

Crust   0
                                                  –30km
Upper mantle   –700

Lower mantle   –2900

Continent 2.9   3.3 – 4.3   4.3 – 5.5   10 – 13.6   Core

                                                  –6371

descending into the mantle. Thus the ocean bed is slowly expanding outwards and this process is known as ocean-floor spreading. The driving mechanism is believed to be a system of convection currents forming cells in the top part of the mantle, as shown in Fig. 2. The rock material in the upper part of the mantle must therefore be in a plastic state, but of extremely high viscosity, and over tens and hundreds of millions of years total movements can be of the order of thousands of kilometres. Regions where the currents of rock in convection cells are diverging form ridges above the average ocean-bed level, for example the Mid-Atlantic Ridge; where they are converging, trenches form, going down to nearly 10 000 m. The Aleutian Trench is one example. The land areas of the continents are also moving, being carried by the convectional movement of material below them. These movements in the outer zones of the Earth explain why some parts of the Earth's surface are rising, some sinking, and some are moving horizontally. Upward movements form mountains, but while they are being formed the surface forces of the sun, wind, and rain attack the rock and decompose it from a solid state, forming loose pieces of rock particles (soil) which are removed by erosion and deposited at lower levels. Weathering is the word used to describe the processes of rock decomposition; erosion is the removal of the weathered rock by rivers, wind, and ice to other places, low ground or into the ocean. Rock material is therefore continually moving and during this movement undergoes changes which will be described in Chapter 2. The movements which form mountains are not continuous constant rate processes, but occur at irregular time intervals. Stresses build up in the rock and cause deformation and increase in strain energy. When the rock reaches the elastic limit a fracture occurs and there is a sudden release of energy in the form of earthquake waves. The fracture in the rock is called a fault. The movement during a single earthquake may be of the order of a centimetre, but in a severe earthquake it may be as much as a metre. Much of the released energy travels away from the fracture as a surface wave, like a wave on the sea, and this is the reason why buildings are destroyed during an earthquake. Some parts of the Earth are called seismic areas because there are

Fig. 2. Convection currents in the mantle form trenches at the margins of the continents and ridges in the oceans.

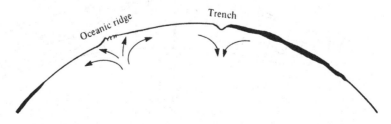

active mountain-building forces in action and earthquakes are common. Other areas are mountainous but the forces which made them have died away and earthquakes are rare, or do not occur. Large areas within the continents are low ground, or are elevated flat areas known as tablelands. Any mountains which were once in these areas have been eroded away and seismic activity has stopped.

All these processes of rock formation, movement and decay have caused rock found at the surface (outcrop) to consist of many different types and to have definite physical properties. The important geotechnical properties are specific gravity, permeability, strength, compressibility, and state of weathering. The last is very important because the other properties are dependent on it. The different properties are also dependent on each other. Measurements of these properties are made at proposed construction sites (*in situ* tests) and on samples taken to a geotechnical laboratory, in the hope that the mechanical behaviour of the rock or soil can be predicted and suitable construction designs made.

The movements described here cause compression and tension forces within large masses of rock. Upward movement causes tension cracks to form in hard rock near to the surface because of the increasing length of arc with greater radius (Fig. 3). Open cracks are in general only seen in the top 10 m of rock, but the rock below this level may contain very many planes of weakness, which only become discontinuities when the rock is near the surface and has the freedom to expand. Relief of compression stress caused by the weight of overlying masses of rock also causes rock when it gets near the top surface to expand radially outwards, forming more discontinuities with a dominantly horizontal direction. Geologists call these fractures joints, but in geotechnical language they are discontinuities. Hard rock in the ground often breaks easily into pieces that have regular geometrical shapes – cubes, rhombohedra – caused by the stresses that have been acting on the rock. Some rocks break into very irregular pieces because in the past they have been affected by several phases of severe stress acting in different directions at different times that have left their mark on the rock.

Fig. 3. Tension cracks forming in rising and expanding rock.

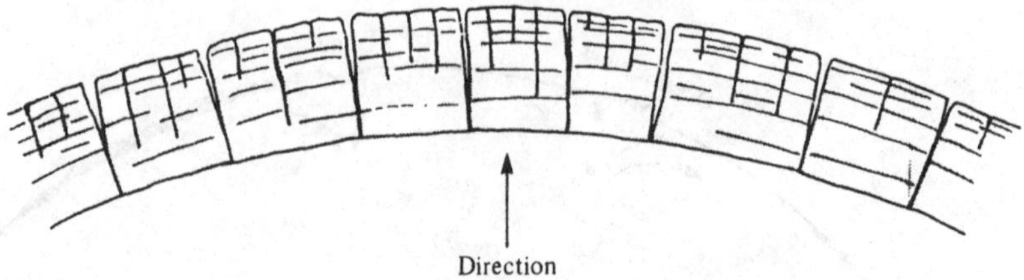

Direction
of stress

These discontinuities are described in the section on engineering description of rocks in Chapter 6. A study of rock exposed in a deep excavation shows that the number of discontinuities per unit volume decreases with depth and the deeper rock appears to be more solid. One exception to this is limestone, which can easily be dissolved by water in the ground and which usually contains many fissures, often large.

Weathering and decay of the Earth's surface

The natural tendency of rock near the surface to break into small pieces causes it to have a greater surface area per unit volume, and water can enter the discontinuities. In humid climates this accelerates the rate of decomposition of rock at the surface. Weathering in humid climates is chiefly a chemical process. Rainwater is very slightly acid because of dissolved carbon dioxide from the atmosphere; plants emit carbon dioxide from their roots, which also manufacture humic acids, and the interaction of these with rock releases solutions of plant nutrients, phosphorus, potassium, calcium, and trace elements, that are necessary to the plant. Water acts as the solvent in these processes by which plants decompose rocks and break them down into soil. The removal of these elements from the body of the rock leaves very small spaces into which water enters and the decay process accelerates. The rock first becomes porous, then breaks into small pieces which become progressively smaller as the rock is changed into soil. The chemical processes are complex and depend on the various minerals of which the rock is composed. Silica (SiO_2) in the form of the mineral quartz is stable in temperate climates, but in equatorial climates the high temperatures help it to decompose. Minerals that contain iron decompose to form iron hydroxides, which give the rock a brown colour, an indicator of weathering when fresh, unaffected rock is not naturally brown. Plant roots enlarge discontinuities by forcing the rock surfaces apart, helping the weathering process.

In the cold climates of high latitudes, and in mountains, the sequence of freeze and thaw of water breaks up rock because of the expansion of water when it freezes. The process may be seasonal, wet rock after the summer may be frozen for the whole winter, one cycle per year, or the process may be partly diurnal when rock exposed to the sun thaws during the day and freezes again at night.

Wind forces may be strong enough to blow away pieces of rock that have become loose because of other weathering processes. The greatest wind erosion effects are seen in dry climates where the wind can blow strongly enough to carry with it tiny pieces of rock, grains of sand or dust. This has an abrasive effect, sand blasting, which carries away large quantities of rock as more dust. Wind erosion produces some spectacular rock shapes in desert climates, for example Monument Valley in the United States of America.

The total effect of all these processes is to build up a mass of loose rock material above a more solid body of rock below (bedrock). The boundary between the two different masses is called rockhead, and is shown in Fig. 4. The top layer is called the weathered or superficial zone, or overburden. The depth of the weathered zone may be very important in construction works. Towards the level of rockhead there is usually a greater proportion of larger pieces of the underlying rock, contained in soil which binds the whole together. The superficial zone material is not always produced in the place in which it is found (*in situ*). The material may have been brought in from another area by rivers, forming alluvial deposits, or by glaciers, forming glacial deposits, or by wind, forming aeolian deposits. The material below rockhead may not be hard rock, but soft. It will have different mechanical properties from those of the weathered rock above.

This superficial zone of broken and partly-decomposed rock is usually that in which civil engineers have to build their constructions. The material found in this zone is described in detail in Chapter 3.

This very simplified account of Earth processes describes the origin of the rock material and the rock structures which the civil engineer will meet during the course of his work on construction sites. It will help towards an understanding of the various forms of rock trouble which can occur during and after construction.

Earth history

The history of the Earth goes back over a very long period of time, of the order of thousands of millions of years, and during that time certain events occurred which left their mark on the structure of the upper parts of the crust, which we see today. This structure includes the present-day distribution of

Fig. 4. Superficial weathered zone above rock which shows expansion joints.

Soil
Rockhead
Bedrock

continents, mountains, relics of old mountains that have been worn down by erosion, unconformities, and the record of the evolution and extinction of certain forms of animal life preserved in the rocks as fossils.

Geological periods and time scale

Geologists divide Earth history into eras, divisions of which are known as periods. The names used for the periods are based on names for rocks, place or area names, and the kinds of fossils found in the rocks. The sequence of all the evolutionary stages in life, and the sequence of rock formation has been compiled by geologists working over more than a hundred years in all parts of the world, firstly in Europe, then in North America, and the other continents. The internationally recognized periods are given in the form of a vertical sequence, the geological column, with the oldest rocks at the bottom (table 1). This is the standard way of presenting rock information. The oldest rocks in the area described are placed at the bottom of the column. This column can be compared with a core of rocks extracted from a borehole, where the earliest formed rocks

Table 1. *Geological periods and time scale. (Based, with permission, on* The Phanerozoic Time Scale. *Geological Society of London 1964)*

Era	Period		Age (million years before present)
Cainozoic	Quaternary	Recent Pleistocene	
			2
	Tertiary	Pliocene Miocene Oligocene Eocene Palaeocene	
			66
Mesozoic	Cretaceous Jurassic Triassic		
			226
Palaeozoic	Permian Carboniferous Devonian Silurian Ordovician Cambrian		
			570
Pre-Cambrian			

are lower down, and the more recently formed rocks are above the older. It is possible during Earth disturbances for whole sequences of rock groups to be turned over and inverted, so it is not an invariable rule that the rock at the bottom of a borehole is older than the rock at the top. The geological column is a model of Earth history in terms of rock groups, which are known as formations. Rocks of all the geological periods are not necessarily found in any area because that area may have been land during a particular period and rock would not have been forming there at that time but would have been in a state of decay and erosion and providing material for the formation of new rock in another area, possibly hundreds of kilometres away.

Study of Earth movements or disturbances, uplift of mountains, or intrusions of great masses of igneous rock under them, shows that the intensities of these movements build up to peaks and then decay and disappear, leaving mountain ranges and relics of volcanoes which are then eroded down to sea level. The geological column given here states the approximate ages of the time boundaries between the periods. The eras are known as Pre-Cambrian, Palaeozoic (ancient lift), Mesozoic (medium life) and Cainozoic (recent life). The term Tertiary originated from an earlier historical arrangement which divided rock formations into Primary, Secondary and Tertiary groups. The Pre-Cambrian era is so named because it pre-dates a group of rocks found in Wales (Cambria) which contain the earliest forms of quite large marine animals. Subsequent research has shown that more primitive forms of life were abundant in rocks older than the Cambrian.

Geological periods are further divided into stages and zones on a basis of the different fossils found in the rocks. These smaller divisions of geological time are not of much importance to the civil engineer unless the rocks themselves have markedly different characteristics from those above and below, which will affect their behaviour during engineering works. These divisions of periods can be seen in geological maps for all areas, but the smaller time divisions have names related to more restricted area distribution and vary from one country to another. The civil engineer may find such terms included in the geological maps of the area in which he is working, but he is more concerned with the rock type and its structure than with the precise age in terms of the fossil animals it contains. The period names are very widely used and may have some geotechnical significance. The oldest rocks, the Pre-Cambrian, may have been subjected to several earth movements, which may have left them in a severely crushed state and full of discontinuities with very irregular spatial distribution patterns. Palaeozoic rocks are generally hard rocks, but from Mesozoic times onwards some rocks (sands and clays) are still in their original unconsolidated state. The younger rocks of the Tertiary and Quaternary periods are very often unconsolidated formations of gravel, sand, and clay, except for those rocks which were cemented during their formation by precipitated cementing material, calcite and silica, which has formed the hard rock types sandstone and limestone. The term **'consolidated'** is used here in a general sense meaning solid rock, distinct from soft rock; another term used is indurated. The term 'consolidated' is used in a special way in the science of soil mechanics. Clay can be **normally consolidated** or **over-consolidated**.

The igneous rocks produced by lava flows from erupting volcanoes harden on cooling and come into the hard rock group. They are found in all ages of rock formation from the Pre-Cambrian to the Recent, and can indeed be seen in the process of formation at the present day.

2   Rock-forming minerals and processes of rock formation and decomposition

Rock-forming minerals

The word mineral is defined as a naturally-formed substance of definite chemical composition found in the ground. Rocks are composed of different types of minerals, mostly compounds of silica (silicates), and carbonates. In the usual sense the word mineral means something extracted from the ground for its value, often metal content. The geological use of this word includes all naturally-formed substances that make up the composition of rocks. Some rocks may be composed of only one type of mineral, others may contain several minerals. Usually about 95% of a rock is composed of three or four minerals, and the remaining 5% may contain as many as 20 types. The major minerals are the important ones from the engineering point of view. There are thousands of mineral types known, but for practical geotechnical purposes only a knowledge of the common rock-forming minerals is necessary. They should be studied from the point of view of chemical and mechanical stability because the strength of a mass of rock is dependent on these properties. The overall strength of any structure is determined by the strength of its weakest members and this principle also holds for rocks. Some minerals, especially those which contain much iron and magnesium and carbonates, decay more quickly when exposed to the atmosphere and water in the ground than others in the rock. Minerals that are stable in humid temperate climates may be unstable in equatorial climates. Their decay reduces the strength of the rock even though other minerals have not decayed. Quartz (SiO_2) is a very common rock-forming mineral and it is highly resistant to decay, except in equatorial climates. The chemical behaviour of minerals in any rock that is used for aggregate in concrete may be very important. There is a chemical reaction, the alkali-aggregate reaction, which damages concrete by the formation of new mineral material which occupies a greater volume than the original material, thus splitting the concrete.

The specific gravity of a rock is controlled by the individual specific gravities of the minerals of which it is composed. The specific gravity (S.G.) of rock-forming minerals ranges from about 2.3 to about 5. Minerals of heavy metals, e.g. cassiterite (SnO_2 – S.G. 6.9), are more dense.

Major rock-forming minerals

The major types of rock-forming minerals are given in table 2. Minerals have definite internal structures built up from the combination of the atoms into groups. The way in which the atoms are built up into groups has some important effect on the mechanical strength of the minerals in rocks. When the structure is in the form of a sheet the mineral can be easily split. Other minerals have an internal structure that only allows them to break into irregular shapes. The carbonate minerals are built from rhombohedral-shaped molecules and a piece of the mineral will tend to break into similar shapes. Because the major rock-forming minerals are silicates (except the carbonates), the molecular structure is described in terms of the structure (lattice) of the silicon and oxygen atoms. The excess ionic charges are balanced by cations of the other rock-forming elements. The chemical composition of some minerals is variable because some elements can substitute for each other in the structure, for example silicon and aluminium, magnesium and iron. This substitution is indicated in the formula thus: (Si, Al), (Mg, Fe).

Olivine (Mg, Fe)$_2$SiO$_4$

Magnesium and iron are interchangeable within the lattice. Light green, darker with higher iron content. S.G. 3.2–3.6. In basic and ultrabasic rocks, e.g. olivine–basalt, peridotite. Easily decomposed.

Pyroxenes

Orthopyroxene: (Mg, Fe)$_2$Si$_2$O$_6$. *Enstatite*: Mg$_2$Si$_2$O$_6$. *Hypersthene* contains approximately equal amounts of magnesium and iron, S.G. 3.2. Monoclinic pyroxene (*augite*): (Ca, Mg, Fe, Al)$_2$(Si, Al)$_2$O$_6$. This is the common form of pyroxene found in basic and ultrabasic igneous rocks. Dark green, it gives these rocks a greenish-grey colour. S.G. 3.3–3.5. *Diopside*: CaMgSi$_2$O$_6$. Light green.

Table 2. *Rock-forming minerals*

Olivine	Separate SiO$_4$ groups	SiO$_4$
Pyroxenes	Single chain	Si$_2$O$_6$
Amphiboles	Double chain	Si$_4$O$_{11}$
Micas	Sheet	Si$_4$O$_{10}$
Feldspars	Framework	(Al, Si)$_n$O$_{2n}$
Quartz	Framework	SiO$_2$
Clays	Layers	Si$_2$O$_5$/X(OH)$_6$
Carbonates	Rhombohedral	XCO$_3$

(X = a possible element)

Amphiboles

Hornblende is the common type: $(Ca, Mg, Fe, Na, Al)_{3-4}(Al, Si)_4O_{11}(OH)$.
Green, common in igneous and metamorphic rocks. Decomposes easily. Crystals
are rod-shaped and in metamorphic rocks are often aligned, as in hornblende-
schist. Asbestos is a form of amphibole.

Micas

Muscovite: $KAl_2[(Si_3Al)O_{10}](OH)_2$. S.G. 2.9. Colourless. Six-sided
flat-shaped mineral. It reflects light well and is often seen sparkling in rocks like
sandstones in which it is found as a detrital mineral. *Biotite*: $K(Mg, Fe)_3$-
$[(Si_3Al)O_{10}](OH)_2$. S.G. 2.8-3.1. Dark brown to black. Micas are common in
granite and mica-schist. Biotite is less stable than muscovite and it is not a mineral
of sedimentary rocks.

Feldspars

Aluminium silicates of potassium, sodium and calcium. There are three
major types: *Albite*: $NaAlSi_3O_8$; *Orthoclase*: $KAlSi_3O_8$; *Anorthite*: $CaAl_2Si_2O_8$.
Albite and orthoclase can form mixtures, known as alkali feldspars or perthite.
They are very common in acid igneous rocks, and make up about 75% of the
mineral composition of granite. White, but may be stained with other colours.
S.G. 2.56-2.76. Albite and anorthite mix to form plagioclase. This is a series of
minerals albite-oligoclase-andesine-labradorite-bytownite-anorthite. Andesine
is in the middle of the series and is composed of about 50% albite, 50% anorthite.
Plagioclase feldspar makes up about 50% of most basic igneous rocks such as
basalt and dolerite. Feldspar crystals split easily along smooth planes and as
a result are easily seen in rocks because of reflection of light from these surfaces.
Feldspathoids are feldspar-like minerals which crystallize from magmas that have
a relatively low silica content. If there is not enough silica to saturate the bases
(magnesium, potassium, sodium, calcium, etc.) feldspathoids will form. *Nepheline*:
$Na(AlSi)O_4$. *Leucite*: $K(AlSi_2)O_6$. S.G. approximately 2.5. These two feldspathoids
are found in lavas and are not stable.

Quartz SiO_2

A hard and chemically-resistant mineral. It cannot be scratched with
a knife blade, the common test for this mineral. It forms beautiful clusters of
crystals in cavities of rocks and occurs in many different colours: many of
these clusters are translucent and are gemstones. A major constituent of igneous,
sedimentary, and metamorphic rocks. During crystallization processes the
elements combine with silica until they are all located in crystal lattices; any silica
left over then forms quartz. Rocks that are rich in quartz are called siliceous.

S.G. 2.65. Quartz is often seen as white veins in rocks that contain it as a detrital mineral, for example sandstones. Because it is relatively hard compared with other minerals and rocks, it is often found as pebbles on coastal beaches. Note that the white mineral forming veins in limestone is usually calcite. A rock that is composed almost entirely of quartz is called quartzite.

Clay minerals

These are built up in layers, like sandwiches. (1) Tetrahedral SiO_4-type layers formed as multiples of Si_2O_5 or $Si_2O_3(OH)_2$. (2) An octahedral layer consisting of a metal ion (aluminium or magnesium) inside a group of six hydroxyls (OH^-) arranged at the corners of an octahedron. These layers can combine in two ways: two-layer units, e.g. kaolinite, composed of alternate silicon and aluminium layers; or three-layer units, e.g. montmorillonite, in which an octahedral layer lies between two tetrahedral layers. S.G. 2.6–2.9.

Clay minerals have properties that are of great importance to geotechnical engineers. Some swell when wet (swelling clays), and shrink when dry. Such clays can cause trouble in foundations of buildings. Other clay minerals can exchange the cations in their structure (base exchange process) and can be used for extracting lime from water supplies. The types of clay minerals in a mass of clay control its mechanical properties, and these can be very variable. Important types are: kaolinite, halloysite, illite, smectite (montmorillonite group), vermiculite, glauconite. *Kaolinite*: $Al_2Si_2O_5[OH]_4$. *Illite*: $KAl_2[AlSi_3]O_{10}[OH]_2$.

Carbonates

Calcite ($CaCO_3$) is the major mineral in limestones, in which it is mixed with clay. Calcite reacts with dilute hydrochloric acid, producing bubbles of CO_2. The acid test is the standard test used to identify limestone. S.G. 2.71. *Dolomite*: $CaCO_3 \cdot MgCO_3$. This mineral is found in dolomitic limestones. S.G. 2.8–2.9. It reacts with warm hydrochloric acid. *Siderite* ($FeCO_3$) is sometimes found combined with clay and is then known as clay ironstone. S.G. 3.7–3.9.

Other minerals commonly found in rocks
Chlorite

$(Mg, Al, Fe)_{12}[(Si, Al)_8O_{20}](OH)_{16}$. A green, flaky mineral like mica and related to micas. S.G. 2.65–3.0. Chlorite is common in slate, schist, hydrothermal veins, and as a decomposition mineral in basic igneous rocks.

Gypsum

$CaSO_4 \cdot 2H_2O$. White and crystalline. It is common in clay deposits that have been formed in arid climates, and may be dispersed through the clay, or

may be in the form of thin beds in the clay formation. *Anhydrite* is $CaSO_4$ and
has no water of crystallization. Both are of economic importance and are extracted
and processed for use in the building industry as plaster and in plasterboard.

Iron minerals

Magnetite. Fe_3O_4. Small black crystals of this mineral are found in
basic igneous rocks, making up about 5% of the mineral composition. When it
is found in large masses, it is extracted as a valuable ore of iron.

Hematite. Fe_2O_3. Bright red when it occurs in powder form in rock and
gives the rock a general pink or reddish colour. In large masses it is almost black,
with a shining lustre, and is then called specularite.

Limonite. $FeO \cdot OH \cdot nH_2O$. A general-purpose name used for hydrated
oxides of iron which are very common in rocks, giving them their colours in
many cases. Limonite forms as a product of decomposition of minerals that
contain iron, particularly olivine, pyroxene, and amphibole. The colours are
brown, yellowish brown, or yellow and when streaks of brownish material are
seen in rocks it is often an indicator that the rock has started to decay. Some
rocks are brown when they are fresh, so this visual test requires experience
before it will be reliable. Brown clays contain limonite.

Pyrite. FeS_2. A brassy-coloured mineral when found as crystals and
known as 'fool's gold', but is black when it occurs dispersed through a rock as
a powder. Pyrite is often the cause of the black colours of rocks, particularly
clays. It forms under chemically-reducing conditions as a precipitate in oxygen-
deficient environments. When exposed to the atmosphere, pyrite will oxidize,
forming iron sulphate, and swelling. This can lead to instability in clay masses.

Chemical analysis of the rock found in the top layers of the Earth's crust
shows that nearly all of it consists of the following elements in order of abundance by weight: oxygen > silica > aluminium > iron > potassium > sodium >
calcium > magnesium > titanium > phosphorus. The first four make up about
80% of the average composition of rocks by weight. The other metals, lead, zinc,
copper, manganese, tin, gold, silver, platinum are on average rare elements and
are only commercially extractable from the ground when they have been
naturally concentrated into mineral deposits. Water is an important concentrat-
ing agent of these metals, particularly when it is hot, beyond boiling point. It
then dissolves metals and transports them to cooler regions in the crust where
the water becomes saturated with respect to the metals and they precipitate out
as hydrothermal mineral veins.

Rock-forming processes

Three distinct processes bring about the assembly of the rock-forming minerals in various proportions to form rocks. These processes are known as igneous, sedimentary, and metamorphic processes.

Igneous rocks are formed by the crystallization of melted silicate material in the Earth's crust, or on the surface after it has erupted from below during volcanic activity. The name is derived from the Latin word for fire, because volcanoes give the appearance of fire, although the effect is caused by incandescent rock at very high temperatures, up to 2000°C, and not by combustion. This melted rock is known as magma and is formed inside the crust and in the upper mantle by the same processes that cause lifting of the crust to form mountain chains. The origin of this heat is still a subject for discussion among geologists. Melted rock can penetrate into regions that are being pushed up into mountains, forming their foundations and cooling and crystallizing to form solid igneous rocks. This process brings about the formation of large masses of igneous rocks, hundreds or thousands of kilometres long, following the lines of mountain chains or ranges. Active mountain chains like those on the western side of North and South America have volcanoes distributed along their length. Volcanoes occur where the magma has managed to force its way to the surface and erupt. The driving force is superheated steam. Some of the magma may remain in liquid form, running out of the volcano as a lava flow; otherwise it may be converted to a powder by the power of exploding steam as the pressure is suddenly released on escape through the top of the volcano. The volcanic dust settles out in layers and can travel hundreds of kilometres from the centre of the eruption. Fine volcanic dust can travel right round the world and will eventually fall to the ground. The record of past volcanic eruptions can be traced by examination of cores of sediments taken from ocean beds.

After many millions of years the mountain-building forces die away and the area becomes quiet but subject to rock decay and erosion until the mountains are worn down to a flat land surface almost to sea level. The eroded material goes to form sediments elsewhere.

Sedimentary rocks are formed under water, in the sea and in lakes, and as deposits of wind-blown sand and dust. When formed under water these sediments consist of variable amounts of two components: (1) detritus – the product of erosion of the land carried to the sea by rivers; this material is mostly quartz, clay, mica, and fragments of rocks; and (2) minerals precipitated from solution in the water, mostly calcium and magnesium carbonates, iron sulphides, oxides, or hydroxides, and silica as quartz. The sediments accumulate in slowly-sinking sedimentary basins, which are very shallow areas of the Earth's surface, covered by water. Sediments also accumulate in the deeper ocean areas, and in the troughs adjacent to some parts of continents. As a general rule coarser-grained material accumulates in areas closer to the land, finer-grained material farther out to sea. Salt water in estuaries causes clay suspended in river water to flocculate and deposit as estuarine mud. Common sedimentary rocks include beds of clay, silt, sand and gravel, which, when they become hard rocks after the filling of voids with cementing material, form shale, siltstone, sandstone, and conglomerate, respectively. Limestone is another very common sedimentary rock. Rock types are described in detail in Chapter 3.

Metamorphic rocks have been affected by heat and pressure after their formation originally as igneous or sedimentary rocks. Even metamorphic rocks themselves may be altered again by further metamorphic processes. The name for these rocks is derived from the Greek for 'change of form'. When pressure alone has been the agent of metamorphism the process is called dynamic metamorphism; this occurs as a result of the high shearing stresses generated within a mass of rock when it is pushed upwards to form mountains. Some heat may be generated as a result of friction. There is very slow recrystallization and alignment of the mineral material in the rock and chlorite and mica are formed, with parallel growth of crystals, so that the rock can be split into flat pieces easily. This process produces the rock called slate. Thermal metamorphism is the result of heating of rock adjacent to an igneous intrusion into earlier formed rocks. Heat released from the crystallizing magma bakes the adjacent rocks and new minerals form. The new rock is usually very hard and is called hornfels.

Dynamo-thermal metamorphism, also known as regional metamorphism, is caused by high temperature and pressure acting on the original rock. The pressures are usually anisotropic, i.e. not of equal magnitude in all directions, and the maximum temperatures reached can be as high as 700°C, not high enough to melt the rock entirely and turn it into a magma. These processes may last for millions of years and cause a gradual transformation of the original rock-forming minerals. The shapes and orientations of the new minerals are such that the mass of rock occupies less space in response to the very high pressures (up to 5000 bars) that have been acting. The minerals are often flat or rod-shaped so that they pack together closely and occupy less volume. This effect can be compared with a pack of bricks in a pallet and the same number of bricks in a loose heap. Rocks formed in this way are called gneiss or schist. They are formed several kilometres down in the crust but can become visible on the surface as the result of uplift and erosion.

Specific gravity of rocks

Specific gravity ranges from 2.6 for rocks rich in silica and therefore containing quartz and feldspar as the major minerals, to 3.4 for basic and ultrabasic rocks which contain less silica but more magnesium and iron. Empty spaces (voids) in rocks, caused by the way in which the rock was originally formed, or by its later decay on the surface, reduce the specific gravity. Voids make a rock mechanically weak so measurement of the specific gravity of a rock and comparison with the value expected for an unweathered specimen of the same type is a guide to its strength. This is one of the standard geotechnical tests on rock samples.

Weathering and erosion

Rock at the surface is in a continuous state of decay. The minerals are decomposed by solution of elements in water which falls as rain and enters the rock through discontinuities. The process is called leaching and the solutions are carried away in rivers and returned to the oceans, which act as stores of chemical substances. This decay of rocks is called weathering and the product is called soil. Rocks that decompose to form quartz and clays produce thick soil zones. Limestone usually has a thin soil above it because if it is mostly composed of calcite, that mineral is removed entirely in solution and very little rock material is left behind. The formation of soil above a weathering rock is a very complicated process and is dependent on the rock type and the climate. For the same type of rock different soils will be formed in different climates.

The loss of material from rock weathering at the surface causes the rock to be mechanically weaker than fresh rock and therefore not so good for building foundations. The weathered zone should be removed when large constructions are being built. The weathered zone may go down to 20 m below the ground level. If the rock in the area has been affected by faulting there may be broken and highly-weathered rock at much greater depths below the surface. The weathering index is a measure of the state of rock weathering, based on visual inspection. There are seven grades from fresh rock to residual soil (table 6).

Removal of soil and rock particles by wind, rivers and ice is called erosion. The eroded material goes to the sea and forms deposits on the sea bed as gravel, sand, silt or clay, forming beds of new sedimentary rock. The weight of material arriving later and accumulating above squeezes the water from the sediment and upwards from the deeper layers so that they eventually harden (become indurated). These beds may continue sinking into the Earth's crust until they reach zones of high temperatures and melt to form magma. They may on the other hand rise to form new mountains of sedimentary rocks. All these processes of rock formation, uplift, erosion and deposition, are stages in a continuing cycle of geological events, and the results have important effects on the behaviour of rock when it is met in civil engineering works.

3 Rock types

The rock types defined and described here are those listed in 'The description of rock masses for engineering purposes' (Anon., 1977). Table 3 defines these rocks on a basis of origin, structure, mineral composition and grain size. The structure of a rock is a description of its bulk appearance. Sedimentary rocks form as layers and are said to have a bedded structure; other rock types may also appear to be in beds because their mineral content varies and is repeated in layers, giving the rock the appearance of a bedded sedimentary rock although in fact it is not. The expansion of rock near the surface with resultant splitting along horizontal planes (Fig. 3) also gives the appearance of bedding, although the rock may be igneous or metamorphic. Metamorphic rocks very often show alignment of minerals and segregation of minerals into bands of different kinds. This is known as foliation and is one of the ways by which geologists are able to recognize metamorphic rocks. A rock that has no easily-seen internal structure of bedding or foliation, but appears to be homogeneous, is called massive. This word is not used in the usual sense meaning very large, but because the rock lacks the internal structures which often cause rocks to split into layers it is often found in large blocks.

The grain size is measured in millimetres and there are three major divisions. Coarse-grained rocks have crystals of more than about 1 mm dimension and it is easy to see that they are composed of minerals of different colours. The grain size of some igneous rocks can be up to several centimetres, and these large crystals may be enclosed in a matrix of smaller crystals, a structure which geologists call porphyritic. The medium-grain size range goes down to $60\,\mu m$, which is the approximate limit of the human eye's ability to detect individual crystals without the aid of a magnifying lens. Fine-grained rocks usually have a dull surface appearance which may be compared with the matt finish used in photography. Igneous rocks that have cooled very slowly deep in the Earth's crust have had time for large crystals to form, in contrast with those formed near the surface or erupted on to the surface, when rapid cooling and crystallization allow only very small crystals to form. These crystals can be seen when the rock is examined under a microscope. After some practice it is possible to state that a rock is in the medium-grain group by looking at it in bright light and seeing the sparkle of reflected light from broken crystal surfaces.

The chemical composition of igneous rocks is used as a basis for classification. There are four groups, acid, intermediate, basic, and ultrabasic. The terms used are related to the amount of SiO_2 in the rock. SiO_2 combines with the other elements to form the silicate minerals and the process is similar to that when an acid is added to a base to form a salt and water. The bases in rock chemistry are the other common elements described above, magnesium, iron, calcium, sodium and potassium. When there is enough SiO_2 to combine with all these the rock is called intermediate, and is then composed mostly of feldspar and some ferro-magnesian mineral, mostly hornblende. If there is more than enough silica to combine with the other elements it crystallizes out as quartz and the rock is called acid, although this name does not mean that it is acid in the usual meaning of the word. Basic igneous rocks contain more magnesium and iron and less SiO_2, but there can be more than enough SiO_2 than is required by the other elements and a little can form quartz, up to about 5% of the total mineral composition. Ultrabasic rocks contain less than 45% SiO_2 and the minerals that form are deficient in SiO_2. Olivine is one of these minerals and is the common mineral in ultrabasic rocks. Because of the high proportion of magnesium and iron the minerals are dark green, there is less than 30% feldspar and the rock has a dark colour. Colour is not a good guide to the chemistry of a rock unless the grain size is medium or coarse. Whilst acid igneous rocks are light coloured when coarse- or medium-grained because of the high proportion of light-coloured minerals, quartz and feldspar, the same chemical material when crystallized very quickly will have a very dark colour; obsidian and pitchstone can be almost black.

Igneous rocks

Granite

Granite is a coarse-grained acid igneous rock composed essentially of the minerals quartz (10–25%), feldspar (60–80%), mica (2–5%), and some minor constituents, often the boron-containing mineral tourmaline. The feldspar is mostly the Na–K-type, orthoclase and perthite, but there is usually some plagioclase, Na–Ca-feldspar. If there is more plagioclase and less Na–K-feldspar the rock grades into a related type, granodiorite. With less quartz, so that the rock is almost entirely composed of feldspar, there is a transition into syenite. The mica in granite may be the colourless but sparkling muscovite, or dark brown to black biotite mica. Tourmaline is blue-black usually, and is found in clusters of needle-shaped crystals, or in veins.

Granite is formed by slow crystallization from a magma, under the mountain ranges that are being lifted up by large movements in the Earth's crust and upper

Table 3. *Rock type classification. Code numbers are based on sequence in columns. (T*

Genetic Group	Detrital Sedimentary		Pyroclastic	Chemi Organi
Usual Structure	Bedded		Bedded	
Composition	Grains of rock, quartz, feldspar and minerals	At least 50% of grains are of carbonate	At least 50% of grains are of fine-grained volcanic rock	

Grain size (mm)

Rudaceous

	Detrital	Carbonate	Pyroclastic	Chemical/Organic
Very coarse-grained — 60	Grains are of rock fragments			
	Rounded grains: Conglomerate (10)	Calcirudite (21)	Rounded grains: Agglomerate (31)	
Coarse-grained — 2	Angular grains: Breccia (11)		Angular grains: Volcanic Breccia (32)	

Saline Rocks
Halite
Anhyd (42)
Gypsu. (43)

Arenaceous — Limestone (undifferentiated) (20)

Medium-grained	Sandstone Grains are mainly mineral fragments. Quartz Sandstone (12): 95% quartz, voids empty or cemented. Arkose (13): 75% quartz, up to 25% feldspar: voids empty or cemented. Argillaceous sandstone (14): 75% quartz, 15% + fine detrital material	Calcarenite (22)	Tuff (33)	

— 0.06 —

			Fine-grained Tuff (34)	Chert

Argillaceous or Lutaceous

Fine-grained — 0.002	Mudstone (15). Shale (16): fissile mudstone. Siltstone (17): 50% fine-grained particles. Claystone (18): 50% very fine-grained particles. Calcareous Mudstone (19)	Calcisiltite (23)		Flint (
Very fine-grained		Calcilutite (24)	Very fine-grained Tuff (35)	Coal (4
				Others

Glassy

n 'The description of rock masses for engineering purposes', *Anon., 1977)*

orphic		Igneous				
d	Massive	Massive				
feldspars, acicular dark ls		Light-coloured minerals are quartz, feldspar, mica and feldspar-like minerals				Dark minerals
		Acid rocks	Intermediate rocks	Basic rocks		Ultrabasic rocks
		Pegmatite (81)				
tite (51)	Hornfels (61)					Pyroxenite (01) and Peridotite (02)
(52) te layers ular and minerals	Marble (62)	Granite (71)	Diorite (82)	Gabbro (92)		
						Serpentine (03)
(53)	Granulite (63)					
	Quartzite (64)	Micro-granite (72)	Micro-diorite (83)	Dolerite (93)		
(54)						
	Amphibolite (65)					
55)						
te (56)		Rhyolite (73)	Andesite (84)	Basalt (94)		
		Obsidian (74) and Pitchstone (85)		Tachylyte (95)		

mantle. Some of the magma may escape to the surface and erupt in volcanoes, forming rhyolite and tuff, the fine-grained equivalents of the coarse-grained granite. The evolution of granite takes many millions of years of heating of rock material and slow crystallization when the heat supply stops. Large crystals have time to form and make a coarse-grained rock. During the later stages of the evolution of granite intrusions small amounts of residual liquid granite penetrate the earlier formed mass as veins, cooling relatively quickly and forming micro-granite, which is sometimes called felsite. This medium-grained relative of granite is made of crystals in the range 0.05–1 mm, but the rock may also con-tain feldspar crystals up to 1 cm in size, and is then said to have a porphyritic structure. Pegmatite is a type of granite in which the quartz and feldspar crystals are very large, often as much as 5 cm long and interlocking in the middle of the vein (Fig. 5).

After uplift and erosion of the mountains their granite core is exposed as outcrops which may be thousands of kilometres long by hundreds wide, elongated masses following the lines of the former mountain ranges. These large granite masses are called batholiths, from the Greek for 'deep rock'. The rock that surrounds the granite is called country rock. Because of the very great forces that accompanied the intrusion, including shear forces, the country rock is found in a very contorted state, folded into tight folds, with many large fractures (faults) and joints, and slaty cleavage may be formed in any fine-grained rocks adjacent to the intrusion. The heat released by the cooling and crystallizing granite magma alters the surrounding rocks by a process of thermal (or contact) metamorphism. New minerals are formed and the country rock may become very hard because of the baking process, when it is called hornfels. The zone of thermal metamorphism may extend radially for many kilometres from the granite margin, or only a few hundred metres, depending on how much heat was released from the magma. Some granite magmas must have been very hot and active because the contact zone with the country rock is not clearly defined, but the two rock types are very much mixed together, forming a rock called migma-tite. Where the contact between granite and country rock is sharp and there is only a narrow zone of thermal metamorphism, the granite cannot have had much energy left in it. The rocks in the outer zones of thermal metamorphism are often spotted; the spots of 1–3 mm size are centres of new mineral growth which ceased to form larger crystals and make a new rock type because the heat supply was not sufficient to complete the transformation. The surrounding zone of thermal metamorphism is called the metamorphic aureole.

Metallic elements such as tin, copper, lead, and zinc, which may be in the granite magma, do not fit into the crystalline structure of the granite minerals (quartz, feldspar, etc.) but remain in solution in the hot water which is escaping

Fig. 5. Granite: (*a*) typical granite batholith; (*b*) vertical section of granite boss; (*c*) pegmatite; (*d*) granite; (*e*) microgranite.

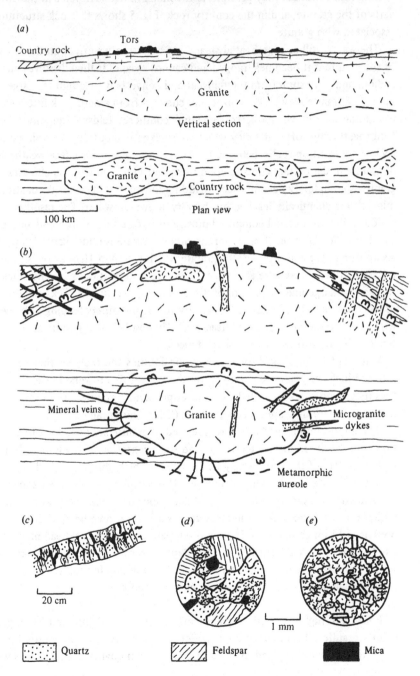

(*a*)

Tors

Country rock

Granite

Vertical section

Granite

Country rock

100 km

Plan view

(*b*)

Mineral veins

Granite

Microgranite dykes

Metamorphic aureole

(*c*)

20 cm

(*d*)

(*e*)

1 mm

Quartz   Feldspar   Mica

from the cooling mass into the surrounding country rock. When these metal-bearing solutions cool they precipitate the metal minerals in veins in the outer parts of the granite, and in the country rock. Fig. 5 shows the rock structures associated with granite.

The water, still as superheated steam, has a destructive effect on the feldspar of the granite if it becomes trapped and cannot escape. It then dissolves out sodium and potassium from the feldspar and converts it to kaolinite. Large masses of granite may be kaolinized and changed from a hard rock into a soft crumbling state in which quartz, some undecomposed feldspar, and mica are held together by soft white clay which can make up about 20% by volume of the rock. Kaolinization is also believed to occur as the result of surface weathering of granite under certain climatic conditions, but in humid tropical climates granite is completely decomposed, including the quartz, because its constituent minerals are chemically leached by the very active rainwater. The result is a reddish-brown material composed mostly of hydroxides of iron and aluminium, which is called laterite. Deep under mountains granite is under very high pressure. As uplift and erosion proceed, the granite which is within 10 m of the surface is released from this pressure and can expand, cracking as it does so. The cracks form a regular geometric pattern of discontinuities within the rock, one set more or less horizontal like bedded sedimentary rocks, the others tending to be vertical with mutually-perpendicular intersections. The result is that granite at the surface tends to break into rectangular-shaped pieces.

Water enters the rock through the discontinuities and rock weathering occurs, so that blocks of granite stand up above the surface like towers. In the British Isles these landforms are called tors. More spectacular granite outcrops are found in Africa, where they are called inselbergs ('island mountains').

The important engineering factors related to granite are the state of kaoliniza-tion, and the distribution, frequency, and the orientation of the joints. Granite within 5–10 m of the surface may be expected to be highly weathered and not good for foundations for large structures. The completely-weathered zone at the top consists of residual boulders of still decomposing rock, mixed with quartz feldspar in the form of sand, and clay that is usually brown because of the presence of iron compounds. In humid temperate climates rockhead may be a few metres below the surface, but the depth can vary. In tropical climates the superficial laterite deposit may be so thick that building foundations may have to be designed as for a clay base, bedrock being too deep for excavation because of cost.

Pockets of highly kaolinized granite may be found at depth, and under good quality granite. This has been the experience with some dams on granite founda-tions in south-west England. Instead of improving in quality with depth as would be expected, the bearing strength of the granite has been found to decrease, and there was no advantage in going deeper with the excavation in the hope of improvement in rock quality. Under these conditions it is necessary to grout the porous rock to make it solid. Although granite is normally considered to be a very strong rock because that is its condition when used for building stone and in aggregate for concrete, one should remember that river valleys tend to be located in the weaker parts of the rock at and just below the surface. The fact that there is a river valley in a particular place may be determined by the presence of weak rock below it. Hence sites for dams, very good in other respects, may automatically have some poor ground conditions in the bottom of the valley although the higher ground at the sides consists of good quality rock on the surface (Fig. 6).

Rhyolite is the fine-grained product of granite magma that has escaped to the surface via a volcano. The liquid rock may emerge as a lava flow, forming a mass of rhyolite when it cools and solidifies, or it may explode because of the force of its content of superheated steam, and be very rapidly converted to dust particles which then fall to the ground and accumulate in beds of tuff; coarser material forms a rock called agglomerate. The rain that descends from the steam cloud above an erupting volcano mixes with the dust and converts it into a mudflow. These rocks are known as pyroclastic rocks and they are very variable in composition, because of the violent and rapidly changing events of volcanic eruptions.

Diorite

Diorite is a coarse-grained intermediate igneous rock composed chiefly of white feldspar, Na–K feldspar and plagioclase, and the green ferromagnesian mineral hornblende. There may be up to 10% quartz present. The rock has a variable colour from greenish-white when it consists mostly of feldspar, to

Fig. 6. Bad ground conditions in granite at dam site.

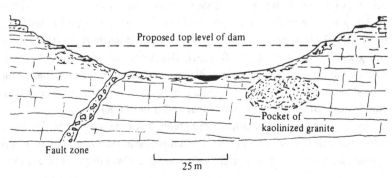

Proposed top level of dam

Pocket of
kaolinized granite

Fault zone

25 m

green when it has a high hornblende content. This rock is a transitional type between acid granite and basic dolerite; it grades into dolerite when it contains more calcium, magnesium or iron, and augite forms as well as hornblende. When there is more sodium, potassium and SiO_2 the rock grades through granodiorite to granite. Microdiorite is the medium-grained variety of diorite and is found in smaller masses such as dykes and sills.

Andesite

Andesite is a common igneous rock produced by volcanoes and gets its name from the Andes mountain range in South America. It has the same general chemical composition as diorite but is fine-grained and may contain some distinct rectangular-shaped plagioclase crystals up to about 3 mm long.

Obsidian and pitchstone

Obsidian and pitchstone are very fine-grained igneous rocks that have solidified immediately after eruption. Both are like dark-coloured or opaque glass and are very hard but brittle. Pitchstone is similar to the slag produced by metal smelting. Obsidian is a natural form of glass and has the same shining surface, but it is dark grey or black.

Gabbro, dolerite, and basalt

These are three igneous rocks in the group called basic because of the higher content of magnesium, calcium or sodium in their composition compared with that of acid igneous rocks. These rocks contain more iron than acid rocks, and it is this element contained in the silicate minerals called ferromagnesian that gives the rocks their green colour. Approximately half of the mineral composition is made up of the ferromagnesian minerals olivine, pyroxene (augite), and hornblende; the other half is plagioclase feldspar, Na-K feldspar may be present, and quartz if there is no olivine in the rock. Oxides of iron and titanium are usually minor constituents, 2-3%. Hence there are several varieties – olivine-gabbro, olivine–dolerite, quartz–dolerite, and many others.

The grain size of these rocks ranges from coarse-grained gabbro with crystals more than 0.5 mm size, through medium-grained dolerite (0.05-0.5 mm) to the fine-grained basalt (less than 0.05 mm). The American name for dolerite is diabase, but in British terminology diabase is the name given to a basic igneous rock that has been affected by thermal metamorphism, with some changes in the composition of the plagioclase feldspar and other minerals.

All these rocks are greenish to grey-green, and dark grey-green. Darker colour values go with higher iron content. Ferromagnesian minerals weather to brown hydroxides of iron and clay, and any brownish colours in the rock are indicators of rock decay. A basic rock that is extremely weathered is a dark brown mass of mineral matter. Joints in the rock are usually lined with the same brown minerals.

These rocks have great strength because of their crystalline structure (Fig. 7e), which consists of randomly-oriented rod-shaped feldspar crystals intergrown with the other minerals, mostly augite, and the whole mass can be compared with glass-fibre-reinforced plastics. Their great strength makes these rocks very good for use as aggregates in concrete.

Gabbro and dolerite occur as sills and dykes (Fig. 7) and the larger masses usually have larger grain size because they have cooled more slowly as a result of their greater mass and therefore greater heat energy stored. The discontinuities are often irregular in density and orientation, in contrast to the often regular arrangement seen in granites. Joint surfaces are irregular and the rock is difficult to excavate, needing blasting (Fig. 8). Some occurrences of dolerite in sills have a distinct columnar structure (Fig. 7*f*), caused by a shrinkage process during cooling of the liquid rock.

The liquid material (magma) comes from the deeper zones in the Earth's crust, escaping to the surface and forming volcanoes by extrusion of lava, which forms basalt.

Basalt is a fine-grained igneous rock, related to dolerite and gabbro, but having reached the surface of the Earth by way of a volcano it has crystallized during rapid cooling, in minutes or hours, and crystals have not had enough time to grow to a size large enough for them to be seen without a microscope. The grain size is defined as less than 0.05 mm, the limit of grain size that can be seen without a microscope. There may be some larger crystals that have crystallized while the magma was below the surface before the eruption. The mineral composition is approximately half pyroxene and half plagioclase feldspar, with up to 5% iron oxide, usually magnetite, Fe_3O_4. Olivine may occur (olivine–basalt) and there may be some quartz, but not with olivine, in extremely small crystals or in the form of glass between crystals of the other minerals. Steam trapped in the liquid rock after eruption expands and forms gas bubbles, which after condensation may react with the hot rock and cause other minerals to grow into the voids, giving the rock a spotted appearance. The texture is then said to be vesicular.

Basalt is found in volcanic cones, and after flowing as a lava down valleys will crystallize and partly fill the valleys. Lava also escapes from long cracks in the Earth's crust, as fissure eruptions, and can then flow out over very large areas because of its great mobility. Basalt from this type of eruption forms sheet structures and in some places can be seen as a layer of very dark rock on top of hills, acting as a protective covering and slowing erosion. Basalt lava can also be erupted by the explosion of superheated steam, which converts the liquid into a dense cloud of droplets. These droplets solidify while falling through the air

Fig. 7. Gabbro–dolerite–basalt series: (*a*) section through volcano; (*b*) sills and dykes invading limestone; (*c*) bedded tuff and agglomerate bands; (*d*) volcanic complex of neck, sills, and dykes; (*e*) microscope section of dolerite, showing augite, rectangular feldspar crystals, and black magnetite; (*f*) columnar jointed basalt.

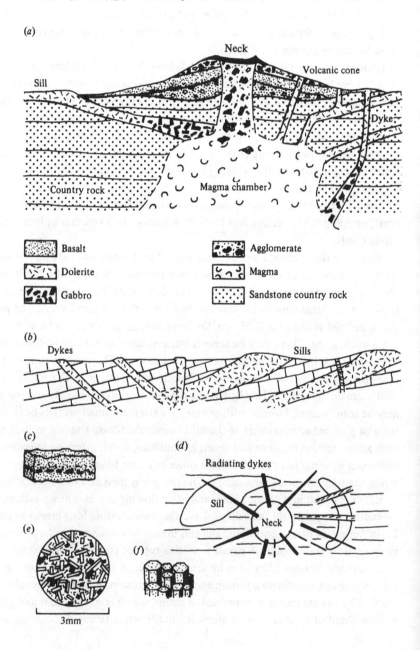

(*a*)

Neck

Volcanic cone

Sill

Dyke

Country rock   Magma chamber

Basalt

Dolerite

Gabbro

Agglomerate

Magma

Sandstone country rock

(*b*)

Dykes   Sills

(*c*)   (*d*)

Radiating dykes

(*e*)   Sill

Neck

(*f*)

3mm

and they build up deposits of volcanic ash. Larger pieces are called volcanic bombs. These air-fall deposits are called tuff when fine-grained, and agglomerate when coarse-grained, as shown in Fig. 7c.

The colour of basalt is usually a dark blackish grey or green, but it may be red or brownish red as a result of oxidation of iron in the minerals to red ferric iron oxide. The texture may be very rough and there may be many voids, which are called vesicles by geologists. These make the rock very permeable and water can easily penetrate and decompose the rock to soil. The weathered basalt is called bole. The reason why people choose to live dangerously close to volcanoes is that the soil formed from these volcanic rocks is very rich in plant nutrients, potassium and phosphorus, and trace elements, and therefore is very fertile.

Contraction of basalt lava during cooling sometimes causes a very distinctive pattern of discontinuities, vertical columns (Fig. 7f) of hexagonal shape. The best known examples are the Giant's Causeway in the north-east of Ireland, and at Fingal's Cave on the island of Staffa in the Hebrides off north-west Scotland.

Basalt is tough and difficult to excavate, requiring blasting. Pockets of highly weathered basalt may be expected to be met during engineering operations. Steep rock faces are usually stable because of the irregular nature of the discontinuities, but there may be falls of small pieces from the face, which can be dangerous to persons below. The composition of the whole rock mass may be very variable because of beds of tuff and agglomerate within the basalt, and there may be large blocks of other types of rock from the vicinity of the volcano's

Fig. 8. Weathered zone above dolerite sill: (a) fresh dolerite; (b) weathered zone; (c) slate; (d) fault zone; (e) minor offshoot from sill; (f) rockhead. The typical irregular joint pattern is shown and the weathered zone grades from loose blocks of dolerite to brown clay at the top. Boreholes to determine depth to rockhead may meet a large boulder (g) which could be mistaken for rockhead. At (h) the dolerite is 5 m below the surface and excavations for building foundations, sewerage, etc. would meet a sudden change from easily-rippable slate to hard dolerite which would need blasting. There is no surface indication of this buried hard rock mass.

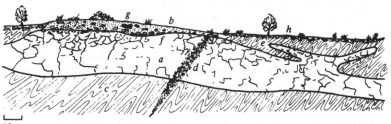

10 m

neck. The geotechnical engineer should expect to find very variable rock conditions, and there can be great variations in the depth of the weathered zone down to rockhead. Fresh basalt (dark greyish green) makes very good aggregate for concrete.

Tachylite is a very fine-grained, almost glassy form of basalt, and is often very vesicular.

Pyroclastic rocks include all those types of volcanic rocks that have been erupted in the form of solid particles or larger pieces, in contrast with those which remained as a liquid mass until crystallized. Tuff and agglomerate are types within this group. Initially they are unconsolidated, but as layer upon layer builds up, the material becomes indurated and forms a hard rock. The coarser pyroclastic rocks are found closer to the volcano, but fine dust can travel very great distances, even all round the world and remain in the upper atmosphere for several years after a very large eruption. Because there are very heavy rainfalls associated with eruptions, coming from the condensed steam, the pyroclastic material near the volcano is often converted into a mudflow, which then flows outwards and eventually solidifies as the water evaporates from the mass. Volcanic mudflows are characterized by great variation in composition. There are often pockets of very rotted rock within only slightly weathered rock. Rock trouble is most likely to occur because of the irregular, non-homogeneous nature and the distribution of voids.

Ultrabasic rocks have a low silica content, less than 45%, and consequently a higher content of magnesium, iron or calcium, which combine with silicon to form the dark-green ferromagnesian minerals olivine, pyroxene, and hornblende as the common types in this rock group. The dark minerals make up more than 70% of the rock, which therefore has a dark green to blackish green colour. The rest of the mineral content is mostly plagioclase rich in calcium and iron oxides, possibly containing titanium.

These rocks form deep in the Earth's crust, slowly crystallizing from material coming from the upper mantle. They can emerge at the surface at a later date after uplift processes. Ultrabasic rocks can also form in thick igneous intrusions by sinking of early crystallized ferromagnesian minerals to the bottom of the cooling mass and forming a concentration.

Pyroxenite is a variety consisting chiefly of pyroxene, usually augite. Peridotite is a rock consisting chiefly of olivine, a mineral which is also called peridot when it is of gemstone quality. Serpentine consists of hydrated magnesium silicates, serpentine and talc. Serpentine is a fibrous mineral related to asbestos; talc is a very soft mineral which feels like soap. The slippery feeling of its surface is caused by its flat-shaped crystals. The serpentine group of rocks is believed to form by reaction of ultrabasic magma with water.

Because ultrabasic rocks are composed chiefly of ferromagnesian minerals which are not stable in surface weathering conditions, they all give rise to geotechnical problems when found close to the surface.

Sedimentary rocks

These rocks can be broadly classified into two groups. One consists of material which has been brought into the area of deposition by water, wind, or ice. This material is called detritus and the rocks made of it are called detrital sedimentary rocks. The other group consists of minerals formed by chemical precipitation from solution in water, or by accumulation of organic residues from plant and animal life. Because the seas and lakes in which these sediments were formed may have had rivers draining into them and carrying detrital sediment, there is often a mixture of detritus and other material formed chemically at the site of deposition and acting as a type of cement binding the detritus into a solid mass. There are therefore many transitional types between detrital sedimentary and the chemically precipitated and organic group.

Saline rocks are precipitates of salts dissolved in water which drains into lakes or inland seas in arid climates. High rates of water evaporation cause the chemistry of the lake to be close to or above the saturation level for a particular salt in water. The salt is then precipitated and sinks to form a deposit on the lake bottom. Thick saline deposits can form when the rate of inflow of water containing the dissolved salts equals the rate of evaporation so that there is a continuous process of precipitation. Halite is sodium chloride (NaCl), commonly called rock salt and used in cooking and for the treatment of highways to remove snow. Anhydrite ($CaSO_4$) and gypsum ($CaSO_4 \cdot 2H_2O$) are sulphates, both used in the chemical and construction industries.

There are many other types of saline rock, all consisting of minerals in the form of carbonates, sulphates, chlorides, and borates, all of which have important uses in chemical industries. Saline deposits are usually found interbedded with fine-grained sediments like clay, marl, and mudstone.

Flint and chert are hydrated forms of silica (SiO_2), formed under water. Chert is found as beds, sometimes as thin beds about 10 cm thick, but flint is in the form of nodules, pieces of very irregular shape (Fig. 9*h*). Flint is found in the Chalk, a carbonate rock of Cretaceous age, but chert is found in all sedimentary rock formations of all ages back to the Pre-Cambrian.

Both these rocks are very hard and cannot be scratched with steel (this property provides the best way of identifying them). They have very little strength under shock forces and flint breaks into small pieces when hit with a geological hammer. Flint is similar to glass when struck with a hammer and breaks with very sharp edges, which is why it was once used as scrapers, arrow heads, and for other

Fig. 9. Sedimentary rock types: (*a*) conglomerate; (*b*) breccia; (*c*) coarse greywacke; (*d*) current-bedded sandstone; (*e*) shale; (*f*) mudstone; (*g*) chert; (*h*) flint; (*i*) grit seen under microscope; (*j*) fossils in limestone.

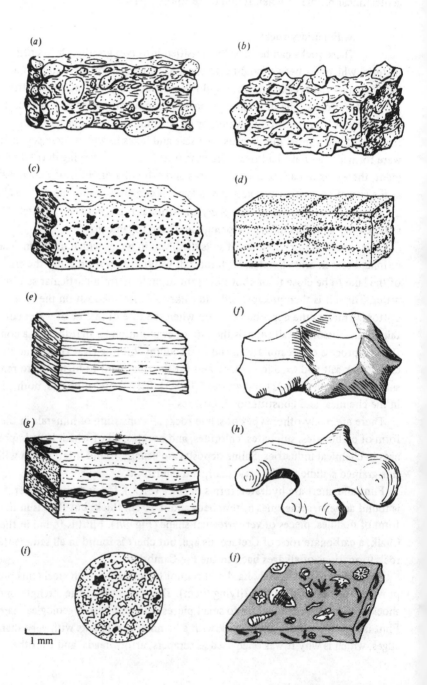

purposes during the Stone Age. Flint is grey to dark grey when fresh, but weathers to a pale brown or whitish skin (patina); chert is often pale brown or whitish, or dark brown to almost black. The colour is very variable and it depends on the chemistry of the water in which the chert bed was formed. The best way of identifying this material is not by the colour but by the scratch test. Bands of chert are shown in Fig. 9g.

Because of their hardness these rocks resist weathering and erosion, and form gravel and coarse sand deposits in river valleys and along coasts, where they may be extracted for use as concrete aggregate. Thin-bedded chert and the nodules of flint can be easily excavated, but thick beds of chert will need to be blasted. Being a hard, brittle, and inflexible rock, chert usually contains many joints, especially when the rock formation has been subjected to folding. The Chalk formation in which flint is found is a soft form of limestone which can be easily broken up by scrapers, when the flint nodules will come out of the rock mass quite easily.

Coal is a deposit of carbon found in layers called seams, interbedded with shale and sandstone in formations of Upper Carboniferous age. This material originally formed as a thick deposit of peat, a mass of plant remains preserved from decomposition by accumulation under stagnant, non-oxidizing chemical conditions, in sinking basins in the Earth's crust. A typical coal deposit contains a mixture of sediments because conditions of sedimentation were changing. After a thick deposit of peat had built up, the sea invaded the area and buried it under a layer of mud; then conditions changed again and sand was deposited. Further changes in sea level allowed the area to become land again and more peat was formed. This sequence of rock formation was repeated many times with the result that coal was formed by dewatering and compression of the whole pile of sediments. When water has been pressed out of a thick bed of peat the thickness is greatly reduced, but the material still contains much hydrocarbon and in this condition it is known as lignite. Further compression drives out the volatile hydrocarbons and converts the lignite into coal. The process can continue, with the assistance of heat in the formation of compressed rock, until about 97% of the mass consists of pure carbon, when it is called anthracite. During the compression, the mud and sand in the original deposit become indurated into shale and sandstone. Because all the beds are thin, of the order of 50 cm, the whole rock mass can be easily excavated. Beds of coal are not economic to extract unless they are at least 1 m thick, unless other factors are present, like absence of any thick coal seams in an area where there is demand, but which would otherwise have to be supplied from deposits thousands of kilometres away.

The types of plants which formed the original peat deposit did not evolve on the Earth until after the Devonian period, so coal is not found in rocks that are older than Carboniferous age.

Conglomerate and breccia are two types of coarse-grained sedimentary rocks. They consist of pebbles cemented together by fine-grained material – clay, silt, sand, quartz, iron oxide (red), or iron hydroxide (brown). The name used depends on the shape of the pebbles, rounded pebbles are found in conglomerate, angular pebbles in breccia (Fig. 9a, b). Many examples of these rocks contain both rounded and angular material, and the word conglomerate is then used, or the alternative rudite, a general-purpose term which covers both types. These rocks are formed by the action of erosion and sedimentation. Rock which has been broken into pieces by weathering has been transported away by water, and as the pieces grind against each other the size of the material is reduced, boulders are reduced to pebbles, then to sand, and during the process any sharp corners in the original pieces are rounded off. Material which has been deposited in thick beds quickly, without prolonged periods of grinding down, remains angular and forms breccia. If there is a long process of rolling and grinding, beds of conglomerate are formed. Pebbles accumulate as beds in river valleys, forming gravel deposits. They are also found on beaches adjacent to hard rock coasts. As these deposits build up in thickness, any water between the pebbles tends to wash down fine-grained rock material, or precipitate minerals which then act as cement and bind the whole mass together to form a hard rock. Various compounds of iron are the common cementing materials. The voids between the pebbles may eventually become completely filled up with cementing minerals and the rock becomes hard and looks like concrete. Because the natural processes are variable in efficiency, being dependent on time and local conditions, there is a continuous sequence from unconsolidated gravel to hard conglomerate. This range of consolidation is important in geotechnical engineering because it controls the bearing strength of the rock, and the permeability, another very important factor. The porosity can be high and the rock can have a high water content and be an aquifer. The porosity often changes laterally as well as vertically from one bed to another, so rapid changes in the groundwater conditions can be expected in excavations and cuttings. These rocks are often quite stable in steep highway cuttings because of the good drainage, which prevents water pressure from building up and breaking the rock mass apart. The coarse material also seems to have a reinforcing effect on the whole mass of the rock formation. These factors are shown in Fig. 10.

Sandstone is an indurated form of sand. It consists of grains of quartz, mica and fragments of fine-grained rocks in the particle size range $60\,\mu$m to $2\,$mm, cemented together by other minerals, often quartz precipitated as a cement. The deposit forms while weathered rock material is washed into the sea or lakes by rivers. While these sedimentation areas are sinking, the deposits build up and the weight of the whole mass squeezes the water in the sediment upwards and compresses the earlier formed layers. There is a progressive dewatering and induration of the initially wet sediment and during this process mineral matter is deposited between the grains and this cements the whole mass until all the voids are filled. The efficiency of the cementing process varies from one part of the rock formation to another so that the porosity of the rock is not the same in all places. The cementing material is usually quartz (SiO_2), calcite ($CaCO_3$), iron oxides and hydroxides, and clay. The grains of detritus may have a small range of particle size, or there may be a wide range from sand size, through silt (4–60 μm) to clay (less than 4 μm). The result of this is that sandstones are variable in mechanical properties, porosity and colour, depending on how they have been formed. There is a continuous range from unconsolidated sand to solid sandstone with almost no voids. The subsequent behaviour of sandstone under weathering conditions will depend on the type of cement; calcite cement is more easily dissolved than silica cement, and the rock is left porous as a result. Arenite is another name for sandstone, and if the grains of quartz are large and angular in shape the rock is often called grit. Calcareous sandstones contain calcite cement; ferruginous sandstone contains iron minerals in the cement. Ferric iron oxide colours the rock red, iron hydroxide makes it brown or yellowish brown. Many sandstones are greenish grey. Slight changes in the chemistry

Fig. 10. Variable grain size in breccia caused by changes in sedimentation conditions. Groundwater emerges at boundaries between permeable and impermeable beds. Small fault at X. Erosion builds small delta fans of rock debris.

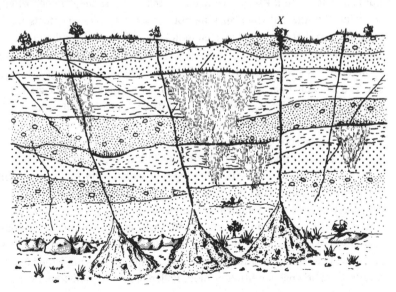

and the grain size of individual layers often give the rock a striped appearance, which is known as lamination and which is a clue that the rock is of a sedimentary type. Sandstones formed under desert conditions have brilliant colours, red, orange and yellow. With decreasing grain size the rock grades into siltstone, and if there is more than 15% clay the rock is called argillaceous sandstone. If both coarse and fine material occur the rock is called greywacke. Almost pure quartz sandstone is called quartzite.

Variations in porosity caused by the degree of cementing are important in geotechnical engineering. Porous sandstones are good aquifers, but the water yield can vary very greatly because of porosity changes. These factors cause variations in the groundwater conditions in the construction site. Porosity changes also affect the bearing strength of the rock, and the stability of slopes cut into it. These variations in the properties of sandstone are found over vertical distances as well as horizontal.

Sandstones in rock exposures often weather so that the bedding is etched out and gives the rock a distinct layered appearance. Thin layers of finer-grained material, which may be clay or shale, will weather back more quickly, accentuating the bedding planes. Any clayey material between beds of sandstone will have an important control on the overall strength of the rock. The bedding planes may be regular in shape, or they may be very variable, curving and intersecting each other, and this structure is called current bedding (Fig. 9d). The cause is a change in the direction of the water currents in the area of deposition. The water in shallow seas is continually moving in different directions, depending on tides and wind, and it keeps stirring up the sediments as they are deposited, with the result that the bedding planes are not flat surfaces. Similar effects are found in sandstones that have been formed in deserts. The curved bedding planes in these rocks are often very noticeable and the structure is called dune bedding.

Mudstone is a fine-grained sediment, originally a mud on the bed of a lake or sea. The individual particles which make up the rock are clay, fine-grained carbon, and iron sulphide. The last two control the colour of the mudstone, which can be almost black when they are present in important quantities. The wet mud on the subsiding sea bed accumulates in layers and the interstitial water contains dissolved materials which later precipitate as calcium and other carbonates, or iron sulphide. As with other types of sedimentary rocks the weight of material above compresses that below and drives the water upwards, so that eventually an indurated sediment is formed. The compression does not cause the formation of a laminated structure as with shale, and when broken the rock has curved surfaces (Fig. 9f). If the iron in the mudstone is in the ferric oxide state the colour will be red. When affected by rainwater in outcrops mudstone quickly reverts to its original condition of mud. Fallen blocks of mudstone tend to expand because of release of pressure, and break into small pieces, giving a larger surface area and greater rate of decomposition. Iron sulphide in a mudstone tends to oxidize and expand; this expansion breaks the rock into very small pieces and accelerates the change to mud. Mudstone often changes in structure to shale (Fig. 9e), a distinctly laminated rock type, and therefore the overall behaviour of the rock mass is variable from one part to another. Mudstone near the surface can usually be easily ripped out in excavations because of its natural tendency to break along curved joint surfaces.

Because mudstone is a fine-grained rock it is impermeable and beds of it between permeable rocks can impede groundwater movement. Water then concentrates at the mudstone boundary, and may cause difficulty in excavations, and be the cause of rock falls and general trouble in the rock faces of cuttings, and in cliffs on the coast. Claystone is a form of mudstone, formed mostly of clay highly compressed and cemented by other minerals, but not having the fissile structure of shale.

Shale is a fine-grained sediment consisting of clay, calcite sometimes, and compounds of iron which give the rock its colour. Shale is formed as a mud in lakes and in seas, where the absence of water currents allows the fine-grained sediments (less than 4 µm size) to build up in thickness. The chemical state of the water determines the type of iron compound or other mineral material being precipitated in the accumulating mud. Stagnant water (oxygen absent) brings about the precipitation of iron sulphide, which colours the mud black. In well-oxygenated waters a lighter coloured mud will form, but the conditions of still water required for the fine sediment to remain in place on the sea or lake bed generally bring about stagnant conditions because of the lack of water currents to bring in fresh supplies of oxygen to the environment of deposition. Dewatering of the mud as the deposit increases in thickness causes induration. Research is still in progress to find the reason why mud can be transformed into two distinctly different rock types, shale and mudstone. It is believed that in shale the clay minerals and very fine-grained mica crystals are oriented parallel with the bedding planes so that the rock splits easily along these directions. In a mudstone the microstructure does not have this alignment of the minerals, so that the rock does not split more easily in one direction than another and therefore breaks into irregularly-shaped pieces. The exact causes of these differences in sedimentation processes are still being investigated by sedimentologists.

All sedimentation processes eventually change, the clay supply to the environment of deposition may decrease because of lower rainfall on the land which provides the sediment, with the result that there is a greater proportion of *in situ* precipitated material in the rock being formed. If calcite is being precipitated in the area the sediment will be a limestone, which can have very variable amounts of clay in its composition. Alternating beds of shale and limestone are very common sedimentary rock assemblages; they are of great geotechnical significance, and are a frequent cause of rock trouble. On the other hand the rainfall may increase, the rivers will then run faster and carry coarse-grained material to the sea so that deposits of silt or sand form, eventually making siltstone and sandstone. The result is great variation in rock type from one bed to another.

Uplift of the sea bed forms land and the beds of shale can be exposed at the surface. The relief of pressure on the beds as they approach the surface causes them to split along the bedding planes in small flat-shaped pieces. This allows water to penetrate more easily, the rock decomposes and reverts to its original condition of mud. Dry shale tends to be impermeable and water moving through a more permeable rock like sandstone, or well-jointed limestone, meets a barrier and tends to concentrate at the junction between the two rock types. On the surface of the ground in humid climates this change in the permeability of the rock below is marked by wet ground and the emergence of groundwater as springs. Even when the surface is temporarily dry because of a short spell of dry weather the generally wet condition of the ground is indicated by the presence of certain types of plants that like wet conditions round their roots and these are indicators of the average water content in the soil.

Shale is decomposed and eroded more quickly than harder rocks like limestone and sandstone with which it is often found. The result is that the ground surface above shale is often concave relative to the convex surfaces of adjacent harder rock outcrops. Alternation of shale with other rocks gives rise to a landscape of alternating valleys and low hills.

The laminated structure and impermeability of shale are the important geotechnical properties. The first means that shale can be easily ripped and excavated without the need for explosives; the second affects the groundwater conditions and surface flow of water at engineering sites. Because it is a solid rock below the weathered surface layer, shale usually makes good foundations for buildings. Beds of shale among other rocks can be a source of trouble. The type of clay mineral in the shale is important; some will swell when wet and cause expansion of the rock mass when it is exposed to rainfall. Below the surface, shale can be relatively dry because of its impermeability; rainfall tends to run off the surface into rivers rather than penetrate into the ground and saturate it. Because of its tendency to weather and quickly change to mud, any shale exposed in excavations is best kept covered, especially during wet weather, until concreting has sealed off the rock from external influences. Beds of shale between limestone beds often cause landslips. Water moves through the joints in the limestone downwards until it meets the shale, which becomes changed to mud. The mud layer acts as a lubricant between beds of limestone, and if these are sloping out into a rock excavation they can easily slide and a landslip results. The slope of the beds, or dip, is a controlling factor in these circumstances. Because water is the immediate cause of the rock troubles which occur in shale the best remedial measure is to keep the site dry by providing surface drains, or drainage tunnels in severe cases. Pumping may also be used to dewater the rock mass.

Clay and marl are not included in the rock type classification scheme for geotechnical purposes because they are not indurated types and therefore come within the field of soil mechanics. They will however be briefly described here because they are very important geotechnical materials. Both are fine-grained unconsolidated sediments formed on the beds of lakes or the sea, first as a mud. This gradually becomes dewatered as the weight of sediment builds up in the subsiding area. The wet mud becomes dry and compacted, but not so thoroughly as when mudstone and shale are formed. Marl is clay mixed with a small amount of lime as carbonate or sulphate. Clay is a group name for a number of different mineral species, described in the section on rock-forming minerals. These clay minerals are built up in the form of extremely small flat-shaped crystals, which can only be seen when magnified in an electron microscope. The clay seen in muddy water is usually in the form of clusters of hundreds of crystals held together by electrostatic forces, when it is called flocculated. These clusters can be broken and dispersed into separate crystals by adding a detergent or special dispersing agent to the water. This is necessary when for the purposes of soil mechanics tests it is necessary to know the size of the crystals rather than the size of clusters of them. Besides the true clay minerals, 'clay' in the general sense of a fine-grained sediment also contains variable amounts of silica, either in the colloidal state, or as very fine silt. Carbon, mica and iron sulphide, oxides and hydroxides may also be present, and cause clay to have many different colours. As the proportion of coarser material increases the clay grades into silt.

Dry clay exposed in excavations, tunnels, and cuttings shows definite joint structures, like hard rocks. These have a greater density of distribution towards ground level, where the clay is under less superficial pressure and tends to expand after being in the overconsolidated state. The behaviour of clays is dealt with in the science of soil mechanics.

Wet clay often causes trouble to civil engineers. During construction, any clay exposed on the site should be kept dry with covers. The movement of heavy machinery and tracked vehicles over a clay site will quickly have a bad effect on the natural drainage in the clay, through joints, and in wet weather the whole site may become waterlogged, causing delays to the work.

Clay beds among permeable beds of silt and sand will impede groundwater flow and concentrate it at junctions between permeable beds above and impermeable beds below. The junction zone becomes saturated with water and where this effect occurs in a cliff or a rock cutting the strength of the rock mass will be greatly reduced and landslips will occur, particularly after periods of wet weather. Such conditions usually cause continual landslipping and the only satisfactory solution is to keep water out of the area by an efficient drainage system. This may be uneconomic, or even impossible, because the groundwater may be coming into the area from considerable distances, in which case expensive retaining walls may be the only practical solution.

Calcareous rocks contain more than 50% calcium carbonate, as defined for the purpose of geotechnics. Magnesium carbonate is often present as well. The rock group includes limestones, and rocks that are transitional to other sedimentary types, conglomerate, sandstone, siltstone, shale. Coarse-grained rocks with carbonate cement are called calcirudite; calcarenite is a carbonate-cemented sandstone, and calcisiltite is similar but finer-grained. Lutite is a general name for any fine-grained sedimentary rock like shale or mudstone, and the name calcilutite is used when there is more than 50% calcite in the mineral composition of the rock.

Limestone is a very important sedimentary rock, used for a great number of purposes, especially for building construction. This rock is usually a hard, massive rock but it often causes trouble to civil engineers for reasons which will be described.

Limestone is composed chiefly of calcite and a smaller proportion of other minerals, including clay, quartz, dolomite, iron sulphide or oxide, carbon, bitumen, and very often shells of sea animals — corals, clams, sea urchins and very many others. Many of the remains of organisms are so small that they require high power magnification before being clearly visible and are called microfossils, but they are very important in determining the ages of some limestones. The sediment accumulates on the beds of warm seas or lakes by precipitation of calcium carbonate together with sedimentation of detrital material, clay, silt, or sand produced by the weathering of rocks on adjacent land areas. Limestones are identified by geologists by the well-known test of application of a drop of dilute hydrochloric acid, which decomposes the calcite and produces carbon dioxide which evolves as gas bubbles. The presence of bitumen or other hydrocarbons can be detected by the characteristic smell given off for a few seconds after a piece of limestone has been broken by the hammer.

There are several varieties of limestone because of the many different minerals which may occur in these rocks: argillaceous limestone (calcilutite) has a high clay content; arenaceous limestone (calcarenite) contains quartz detritus; shelly (fossiliferous) limestone contains abundant sea shells (Fig. 9*j*); dolomitic limestone has dolomite crystals intergrown with calcite; and oolitic limestone or oolite has a texture like that of fish roe.

Much limestone is white when fresh, and when used for buildings produces a characteristic appearance in the architecture of a town in which it is widely used. Dark limestones get their colour from iron sulphide or bitumen in their composition, but they usually weather to white because of the growth of a weathered surface of white calcium sulphate (gypsum), especially in towns where acids in the atmosphere assist the process. Most limestones come into the hard rock classification, but oolite is a softer variety which can be easily cut with saws and therefore makes a good material for the construction of walls, and is known as freestone. Weathering is by solution of calcite by acids in percolating rainwater, which is very slightly acid because of dissolved carbon dioxide from the atmosphere or soil gas, and humic acids produced by plant roots. The calcium is leached from the rock as the bicarbonate dissolved in the groundwater and is eventually returned to the sea by rivers. An insoluble residue collects on the surface of the weathered limestone as deposits of clay and iron compounds, or aluminium hydroxide (bauxite) in tropical climates. Because of the powerful capacity to dissolve limestone, rainwater falling on limestone quickly disappears below the surface via solution channels, pot-holes and caves, which are progressively enlarged by the groundwater. Lack of surface water in streams and rivers is a characteristic feature of limestone terrain. The underground water channels are irregular in size, shape, and distribution, and often cause trouble during construction work because they are not easily and definitely found in advance of the beginning of the work. Excavation may break into large caves, which will have to be filled, adding to the cost of the job. There may be cavities just below the foundations of a construction, undetected until collapse occurs later, possibly under a highway as the result of vibration caused by heavy traffic. Dams for reservoirs on limestone may experience trouble after a few years of service because of solution of the rock by reservoir water seeping into the bedrock because of the head of pressure. Unless the bed of the reservoir is made impermeable by the application of grout, leakage under the dam can become so great that water is no longer impounded.

Beds of shale are often found between beds of limestone. The shale returns to its original condition of mud and then acts as a lubricant at the base of any blocks of limestone that may be sloping out into a cutting in the rock, or from a cliff. This is a common cause of rock slides in limestone terrain. Most limestones contain insoluble clay, which remains on solution and fault surfaces as a slippery mud, helping blocks to slide, especially during wet seasons of the year. Limestone beds are hard and rigid masses, and therefore when the rock has been strained by distortion of the Earth's crust it breaks and is then seen to be crossed by many joints, which are often at right angles to the bedding. This discontinuity pattern is very characteristic of limestone, which is the reason why it is used symbolically for representation of limestone beds in drawings of rock formations. Solution of limestone by rainwater is the cause of the karst type of limestone landform. Instead of broad river valleys there are deep vertical sides to valleys, which only contain permanent rivers if erosion has reached another layer of an impermeable rock like shale below the limestone. In karst terrain any rainfall quickly disappears underground via the joints and caves. When erosion has gone on for a very long time only isolated blocks of limestone are left in the landscape, often very prominent because of their vertical sides, standing up like towers. Karst terrain has a very irregular top surface as the result of these factors. Eventually all the limestone may be removed and another type of landscape evolved on another type of rock below the limestone, but the process may be stopped by a change in the climate, for example to arid conditions. If this happens the different processes of weathering and erosion can cover the former karst terrain with desert type deposits and bury it completely. Excavations into rock that has buried karst may experience trouble when soft rock suddenly changes into a hard limestone when the buried karst is exposed. The hard limestone will be more expensive to excavate and much trouble can be caused by vertical transitions from hard to soft rock.

Although limestone is a hard rock and may therefore be thought of as a rock type that should give little trouble, civil engineering experience with limestone shows that in fact this rock can be a cause of considerable rock trouble.

Metamorphic rocks

Metamorphic rocks are formed from other types of rock by the action of heat and pressure, either separately, or acting together. A few typical metamorphic rock types are shown in Fig. 11. They have characteristic internal structures caused by alignment of the minerals in response to anisotropic stresses while the rock was being altered from its previous condition. The chemical composition of the rock has not been changed by the metamorphism, but the elements recombine to form new minerals, often with the result that the whole mass occupies less volume in response to the very high pressures.

There are two groups of metamorphic rock: (1) foliated types, in which the minerals are oriented in certain directions rather than randomly, and sometimes are segregated into groups of different kinds so that the rock has a striped appearance; gneiss is an example of this, consisting of segregated bands of quartz, light coloured feldspar, and dark mica; and (2) massive types, having randomly oriented (anisotropic) texture, the result of the rock material having been affected by uniform pressure fields rather than shearing stresses.

Fig. 11. Metamorphic rock types: (*a*) gneiss, with dark bands of biotite mica; (*b*) garnet–mica–schist; (*c*) slate; (*d*) mineral lineation in hornblende-schist (amphibolite); (*e*) quartzite. Interlocking grains of recrystallized quartz; (*f*) mylonite. Finely-ground indurated rock particles; (*g*) transformation of clay into slate: (*i*) randomly-oriented clay particles; (*ii*) shear forces generated by folding of beds; (*iii*) cleavage parallel to axial plane (*x–y*) of fold.

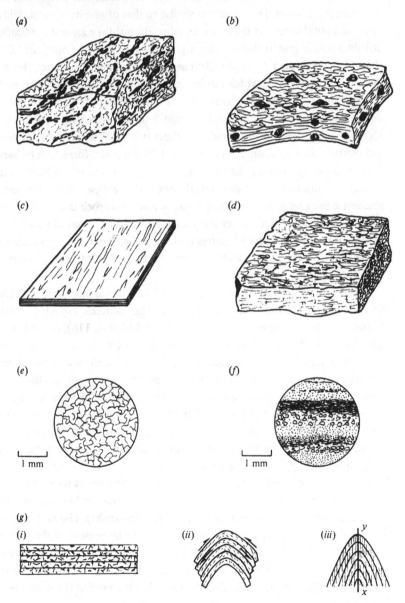

Metamorphic rocks have similar chemical and mineral composition to that of the various types of igneous rock, but the minerals are often arranged in orientation patterns which give a clue to the origin of the rock as metamorphic and not igneous. The rock often has a definite grain, like wood, and this can have some control on the mechanical properties of the rock mass, in some cases very important control, for example the cleavage in slate determines the strength of the rock.

Gneiss has a mineral composition similar to that of granite: quartz, feldspar, mica, and hornblende are the common minerals, but they are not randomly distributed as in granite, being often segregated into bands of one mineral type (Fig. 11*a*), when the rock is described as having a foliated structure. This gives the rock a distinct grain which can influence the properties of the rock mass.

Migmatite is formed by mixture of rock material of different chemical types during metamorphism, for example silica-rich or acid rock with basic rock rich in Mg and Fe. The result of the heating of these two rocks without complete melting and mixture the two components, is recrystallization and formation of bands of dark minerals representing the basic component, hornblende and biotite mica, alternating with zones consisting mostly of quartz, feldspar, and colourless muscovite mica representing the acid component. This rock is a form of gneiss, but the banding is much coarser and individual concentrations of minerals can be of dimensions of the order of metres rather than centimetres as in a foliated gneiss. These different light and dark zones can be seen clearly from a distance from the rock face.

Schist is a metamorphic rock consisting of flat-shaped mica or green chlorite crystals, rod-shaped hornblende crystals and other minerals, especially quartz. Garnet is a common mineral in some types of schist (Fig. 11*b*), occurring as small spherical crystals. The flat- and rod-shaped crystals are aligned so that the rock splits easily into flat- or bar-shaped pieces. The mineral alignment has occurred at temperatures up to 700°C in a strongly anisotropic stress field. Minerals grow more easily along the directions of least stress, and the resulting alignment causes planes of lower strength to form, and the rock is called fissile, breaking into definite shapes, flat, rectangular, or bar shapes. In contrast to slate, a rock formed under similar conditions but at lower temperatures, the cleavage surfaces are rougher because of the larger crystals making up the rock. Schists are formed from clayey rocks, shales, etc., and granitic and basic igneous rocks. Pure clay material is altered to mica–schist, but if there are impurities like magnesium and iron in the clay these will form garnet (garnet–mica–schist). The fact that the garnets occur as separated crystals shows that some movement of the impurity material has taken place; it has been drawn to centres of garnet crystallization spaced as much as a centimetre apart. The crystals of mica cause the rock to sparkle by reflecting light. Hornblende–schist is a type which is formed by metamorphism of a basic igneous rock like dolerite or basalt (Fig. 11*d*). As in gneiss, new minerals formed may separate into bands of different types of mineral composition, giving the rock a striped appearance.

The engineering properties of schist are controlled by the mineral composition and alignment. There may be variations in the strength of a mass of the rock because of variation in the type of schist from one part of it to another. Hornblende may be expected to weather faster than aggregates of quartz and mica.

Phyllite and slate are similar rocks, consisting mostly of muscovite mica and chlorite, with a little quartz. Chlorite is a green, flaky mineral similar to mica, but forming at lower temperatures (up to about 400°C) and in highly anisotropic stress fields. Chlorite is formed when the original material contained magnesium and iron impurities, and was not a pure clay. The circumstances leading to the different types of slate are similar to those causing varieties of schist, and slate is a metamorphic rock that has not been subjected to such high temperatures as schist during its formation. Smaller crystals form at the lower temperatures and possibly shorter duration of the metamorphic process and these rocks are fine-grained and will split easily into flat pieces. When very hard and with flat cleavage planes so that the rock can be split into smooth flat pieces with a broad chisel, the material can be used for the roofs of buildings (Fig. 11*c*). Phyllite is a type of slate which has a texture consisting of flat oval-shaped crystals like tree leaves, which give the cleaved surfaces a characteristic texture. The parallel cleavage planes of slates are caused by the stresses generated when beds of fine-grained sediments are tightly folded (Fig. 11*g*). These planes are intersected by joints which often have a regular geometric pattern with the result that the rock breaks into definite rhombohedral or rectangular shapes. If the frequency of jointing is high, the rock cannot be extracted from a rock face in pieces large enough for making roofing slate. The waste material in a slate quarry is often crushed to small pieces for surface treatment of bitumen-base roofing material to give a 'mineralized finish', or it can be ground to fine dust and then used as a filler in a number of products made of plastics.

Slate is a hard rock, resistant to weathering, which is why it is used on roofs. The orientations of the cleavage and joint systems are the most important geotechnical factors. They determine the way in which the rock mass will break up when being excavated, and the stability of the rock faces in cuttings and other excavations. The stability of a rock face is very much dependent on the angle that it makes with the cleavage and joint system. These factors are described in the next chapter. When close to the surface of the ground slate breaks into a large number of small pieces because of relief of stress. The weathered zone consists of a mass of small pieces of slate in a matrix of clay, the original material from which it was formed and to which it returns during decomposition. There is often a gradual transition from a clay soil to bedrock via a continuous series of changes which show an increasing proportion of larger pieces of slate and a decreasing amount of clay matrix. The weathered zone usually has variable depth because of local factors and on many construction sites the depth to rockhead will need to be found by drilling or trenching during the site investigation. Slate can usually be excavated by ripping, but if the excavation goes deep some blasting may be necessary. Most excavations for buildings will normally go down to rockhead only, unless the level requires deeper excavation, and because slate has a high bearing strength only the weathered overburden needs to be removed. If this is too thick for economic removal, the weathered material can be tested by the usual soil mechanics methods and appropriate foundations designed. If the construction work requires the exposure of a rock face, as in highway cuttings, the site investigation programme will have to include measurement of the orientations and densities of all the discontinuities caused by the cleavage and jointing. Proper design of the slope of cuttings cannot proceed until these factors are known. They are not easy to determine before major excavations are made; cores from boreholes become rotated during drilling and extraction of the drill rods so that the original orientation of the discontinuities in the core sample obtained is lost. There are methods of obtaining oriented cores, but these are more expensive than ordinary methods.

Mylonite (Fig. 11*f*) is an example of extreme dynamic metamorphism. The rock has been crushed along major fault zones between parts of the Earth's crust moving in opposite directions. The forces generate some heat by friction, but the material is not affected by heat rising from some deep source as in the case of schist and gneiss. The rock is fine-grained and has no definite internal structure, but may be banded because of small differences in the composition of the rock which has been ground to dust and indurated. Any type of rock can be converted to mylonite, a word adapted from the Greek for a mill. The process is very similar to milling between millstones, but takes place over a period of millions of years while there are movements along active fault zones in the Earth's crust.

Superficial deposits

Superficial deposits include all the unconsolidated material formed by the various processes of weathering and erosion, which is now found on the surface, or on the sea bed in the broader definition. These deposits are distinct from those unconsolidated materials like clay which have been formed much earlier in Earth history, and they may be considered as being of Quaternary and Recent age. Geology surveyors usually include natural deposits only, and omit man-made deposits from their work. A geotechnical survey will have to include any man-made deposits such as waste tipped into disused quarries or other hollows in the ground. Waste is also used to build up levels during construction and when a site is being redeveloped such material, known as made-ground, will have to be included in the survey. Towns grow upwards as well as outwards during the course of centuries and the waste tip areas of one century become the construction sites of the next.

The processes of weathering and erosion which have been described above cause the surface movement of rock material in a wide range of sizes from metre-size boulders to the finest rock dust. Superficial deposits may have been formed at any time during the last two million years and are of comparatively recent origin compared with the other rock materials. Many of these deposits are still in the process of formation today, but others will be undergoing erosion. The material on the surface may have been produced by weathering of the rock immediately below the surface layer, or it may have come from somewhere else, having been moved by wind, water, or moving ice. The older rock formations below the superficial deposits are called 'solid' by British geologists, but this word does not imply that the rocks are solid in the usual sense of the word, only that they are much older; they may consist of clay and sand, both unconsolidated rock types. It is the age difference that is important in the definition. 'Drift' is another British geological term which is used instead of the term superficial deposit. Nearly all superficial deposits are unconsolidated, except in those cases where there has been precipitation of minerals like silica (siliceous sinter) from hot springs, and surface formation of crusts of hard materials consisting of calcite, saline deposits in some climates in which there are long dry seasons. Any hard material formed on the surface in this way is known as duricrust. The boundary between a superficial deposit and any hard rock below is known as rockhead. Figs 4 and 8 show typical superficial deposits.

The important factors in superficial deposits in the context of geotechnical engineering are: (1) great range of particle size in the material. This controls the permeability of the superficial layer, the stability of slopes cut into it, and the bearing strength of the ground below buildings. The particle size distribution, as defined by a cumulative percentage curve, can vary widely, both vertically and horizontally; (2) variations in areal extent. One type of deposit may be continuous over a large area, tens of square kilometres for example, or it may be in the form of small isolated patches; and (3) variations in thickness.

Most site investigations will be in this sort of material, and the above factors will need to be investigated. The surface between the superficial deposit and the bedrock is important; it may be flat, highly irregular, or very steep to vertical.

There are very many types of superficial deposit; the following are common examples.

Alluvium

This includes deposits of clay, silt, sand, gravel and boulders which are found in the bottoms of river valleys and on their sides (Fig. 12). The material is the product of the weathering of rocks, followed by erosion and transport. The alluvial material is on its way to the sea, very slowly. When climates change and become wetter, rivers run faster and erode alluvium, taking it down to the sea or into lakes. If the climate becomes drier the material tends to accumulate in the valley and increase the thickness of the deposit. A large river valley several kilometres across and hundreds of kilometres long may have been in existence for several million years and have experienced great climatic changes ranging from desert to humid, or even glacial. Variations in sea level may have been of the order of ±100 m during this long period, and this factor controls the cross-section profile and the longitudinal profile of the valley. The Grand Canyon in the United States of America, and the valley of the River Amazon are spectacular examples of valleys. The broad, flat area close to sea level is called the flood plain. It is close to mean sea level at the river estuary, or a few metres above. Careful survey of levels across a river valley often shows that the profile is not a smooth curve, nor linear, but is stepped (Fig. 12a). The flat areas on the sides of the valley are called terraces, or benches. There may be several and they may be given identifying names, numbers, or elevations relative to datum.

The nature of alluvial deposits is very important to civil engineers. River valleys have always been important areas of settlement for man. The deposits range from clay to boulders, and generally one can say that the further downstream from the source of the river the finer-grained the sediment, but there may be important exceptions, for example when a river valley has been occupied by a glacier which has melted and left behind large quantities of rock debris up to boulder size carried along on the way from its source in an upland area. Some construction work will require the depth to rockhead to be determined. This can be very variable and some valleys have great thicknesses of alluvium because they have been 'drowned' by rising sea level and this has caused the sedimentation to rise to levels tens of metres above the base of the original valley bottom related to the former lower sea level. Particle-size analysis of the alluvium is also needed, particularly as the grain size of the material controls the permeability, and groundwater conditions are very important to civil engineering works in river valleys. There is often a large quantity of invisible water moving downstream in the alluvium as well as that visible in the river channel itself. The course of this sub-surface flow of water is determined by the permeability of the alluvium. When an excavation is made into the alluvium of a valley, the water will rise in it, almost up to ground level. This indicates the water table, the surface below which the ground is saturated with water. During periods of wet weather this

Fig. 12. River valleys. Vertical scale 10 times horizontal scale. (*a*) Profile across a river valley. (*b*) Alluvial deposit details. Lenses of gravel and sand in clayey silt. The lens at *x* is a section across a buried former channel of the river. Partly natural, partly man-made banks at sides of river. (*c*) River meander loops in flood plain. Ox-bow lake at (*y*) is part of former meander that has moved downstream. (*d*) River undercutting bank on the outside of meander loop. Gravel piles up on slip-off slope on opposite side of the river channel. (*e*) Longitudinal valley profile. Former sea levels marked 1 and 2.

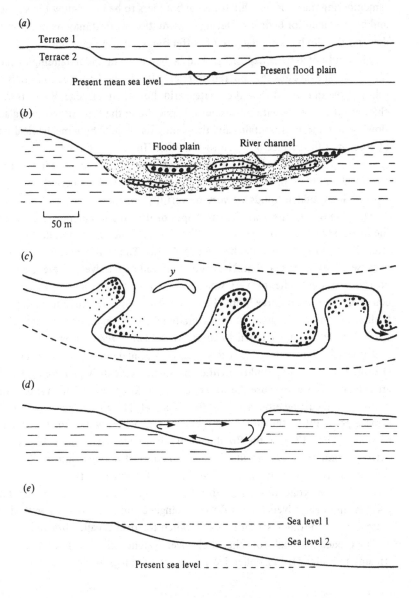

water surface will rise above the ground surface and large areas of surface water will form in the flood plain, without actual overflow from the river bank itself. The water rises from below, through the permeable alluvium. These flooded areas can indicate the more permeable areas in the valley, and if a record is kept of them they will provide useful information for a preliminary survey of the area in connection with some construction project. Knowledge of depth to bedrock is needed if pile foundations are to be designed. This is found by trial drilling, remembering that it is possible for large boulders to be met below the surface and be mistaken for bedrock. During construction work it may be necessary to dewater the site by pumping, use of coffer dams, or grout curtains.

The study of alluvial deposits should be based on an understanding of the fundamental processes that operate in river valleys. The first fact to note is that valleys exist in many different climates, arid, humid, and glacial. Water is the chief agent in the formation of valleys. It may be in the state of ice, as a glacier flowing down from mountains and deepening its channel by a mechanical process of erosion at the surface between rock and ice. In valleys that have permanent rivers the erosion and sedimentation processes are continuous, but in an arid climate the rainfall may be seasonal and very heavy for short periods during storms. Such sudden floods are very powerfully erosive.

The plan of a river shows various shapes of the present channel occupied by the river. Near the source of the river the channel is in a narrow valley and the river alters course rapidly and has angular bends. The water velocity is high because of the steep gradient. Near the other end of the river, where it discharges into the sea, the gradient is very small and the river follows a very winding course in the form of large loops (Fig. 12c). The loops are called meanders and their radii can be very great when the gradient is very small, of the order of 1:1000. Meanders move downstream slowly and an aerial photograph of a river usually shows the former positions of meanders that have moved downstream. These positions will be marked by small crescent-shaped ponds, which are called ox-bows. Because the water velocity is low there is only a small amount of downward cutting erosion of the river channel. The erosive action of the river is more powerful on the outside of the meanders, where water flows with a helical movement (Fig. 12d). In the floodplain the river widens its channel by a sideways cutting action and forms a wide flat area, in contrast to conditions in the region near the source, where the more rapid flow erodes downwards more than sideways. The small gradient causes a river to overflow its channel during high rainfall periods, and many rivers do not flow in a single channel but occupy several, and these are in a continual state of change, when the river is called braided.

The eroded material from the whole river system including its tributary streams and minor rivers, collectively called the drainage basin, is sorted into

different size categories by the action of moving water. The greater the velocity
of the water the greater its carrying power, and the greater the size of the particle
that can be carried in suspension by the water. The mathematical relation
between the size of particle carried and its velocity is a power function with an
index of approximately 6. Doubling the velocity increases the size of particle
that can be carried by about 60 times. The result of this is a very great increase
in the erosive and transporting capacity of a river in times of flood. During its
normal state a river appears to have no erosive ability at all and it does not seem
possible that the river could ever have eroded its valley. Conditions are very different when there is a flood, and boulders of up to 3 m size are moved quite easily.

The result of these factors is that alluvial deposits show great variations in
grain size, areal extent, and thickness (Fig. 12b). The material accumulates as
flat- or lens-shaped masses of different grain size. Geological surveyors record
the area occupied by alluvium as that area with a flat surface between the river
channel and the sides of the valley. The junction between the alluvium and
bedrock is marked by the sudden change in the slope of the ground, and this is
usually taken as the boundary when time does not permit more accurate survey
using augers to sample the material below the top surface. The material is recorded
on the geological map as alluvium without distinction between grain size, except
when there are well-defined patches of gravel.

The information that the civil engineer needs about the nature of the alluvium
includes the depth to rockhead, which can be very variable, generally greatest in
the centre of the valley, but not necessarily so; and the areal extent of material
classified by particle-size analysis. Samples are taken by auger for laboratory
testing. Particle size determines the groundwater conditions in the valley, and it is
also necessary to have advance information from the site survey about the depth
of the water table, the top surface of saturated ground in the valley. The water
table will rise and fall, following river flow changes. There is usually a large
quantity of water flowing through the alluvium to the sea in addition to that on
the surface and it can be very troublesome in excavations. It may concentrate in
buried river channels, former channels of the river which have become buried by
subsequent deposits of alluvium. Buried river channels can be quite deep, tens of
metres below the surface and can have a potential for causing trouble when
tunnels are made across the valley instead of bridges. In cases where underground
working is necessary it is common practice to keep the water out of the working
area by pumping in compressed air.

Glacial deposits

These are also called glacial drift and are superficial deposits of material
eroded and transported by ice, either as glaciers (rivers of ice) or as broad sheets

spreading out from the base of mountains and covering large areas of low ground. The broken up rock material carried in glaciers is called moraine. When the ice melts at the end of the ice sheet or glacier it leaves this material in large irregular heaps, which are called terminal moraines. These may be tens of metres thick, and hundreds of square kilometres in area. The material in the melting end of the glacier is moved and sorted into various particle sizes by the action of water currents. Deposits of gravel, sand, silt and clay are formed, and are sometimes distinctly stratified into layers of different particle size which are called varves. These layers are the result of the seasonal changes in temperature in glacial regions with winter freezing and summer thaw which produce variations in flow velocity of the water from the melting ice, and in consequence repetitive sequences of particle size in the layers.

Specific names are given to the different kinds of piles of glacial deposit: eskers and kames are long ridges; drumlins are semi-elliptical masses which produce a landscape of small, low hills (Fig. 13d). The science of glacial deposits has very many names for a great variety of landforms seen in glaciated areas, but the civil engineer should note only that there are two major groups: those relating to the glacially-eroded rock in the mountainous areas from which glaciers flow, and those relating to the shapes and extent of the material deposited by the melting ice. Only the second group come into this account of superficial deposits, but it is worth mentioning here that the deep channels eroded by glaciers in mountain areas provide very good sites for reservoirs later, when the ice has all melted after climatic changes to warmer conditions.

Boulder clay is a common and important glacial deposit, which has not been sorted by moving water but which has a wide range of size distribution, giving a flat cumulative percentage curve. The range is from boulders to clay. Whereas the sand and gravel deposits left by glaciers may be important sources of con-structional material, boulder clay will have to be washed and sieved into different grades before it can be used. Boulder clay can also be reworked by the river which occupies the valley of a former glacier, and useful gravel deposits formed. The size of the material in such deposits may seem to contradict statements made above concerning grain size and distance from river source, but the explanation comes from the change in climate. Big river valleys have very long histories, which outlast several periods of climatic change.

Head

This is a general-purpose name for weathered rock accumulated *in situ*, not moved after it has formed. It has a wide range of particle size, from clay to boulders. There is a higher proportion of the coarsest material near the junction with the slightly-weathered rocks below rockhead. The depth of these deposits is

Fig. 13. Glacial erosion and deposition: (*a*) lake in trough at head of glaciated valley, cirques; (*b*) terminal moraine; (*c*) lakes on ice-eroded rock surface; (*d*) drumlins; (*e*) eskers.

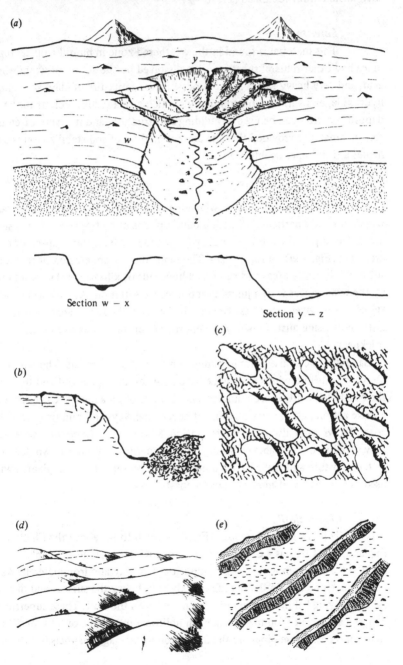

very variable, and when auger tests are made to find the depth to rockhead it is
necessary to remember that the auger may meet a large piece of rock which is
then mistaken for rockhead (Fig. 8 - *g*).

Laterite

This material is found in the weathered zone in humid tropical regions.
Silica and other mineral substances are dissolved by the action of warm, slightly
acid, and very heavy rainfall. The process leaves a mixture of clay, aluminium
hydroxide, and various iron compounds which give a reddish colour to the
deposit. Laterite can be used in building construction, and it is very often used
for road making. Provided that the correct method of soil stabilization is used it
makes a firm impermeable road.

Blown sand

This forms sand dunes, which have a great variety of shapes. Prevailing
winds from one particular direction build up tens of metres of sand in desert
areas. The top surface of these piles of sand may have regular shapes, for example
crescents (Fig. 14*a*), or long ridges. The sand comes from erosion of rock on the
surface in hot, dry areas and can travel hundreds of kilometres. Dry sand blown
by the wind moves with a jumping action called saltation. At a critical wind
velocity for a given size of sand grain, the layer on the surface can be lifted as
a storm of rising dust. Blown sand deposits are unstable masses of variable
thickness.

Sand dunes can also form in humid climates, close to coasts. The sand is
blown by onshore winds from wide beaches when the tide is out and the sand on
the beach dries out. It accumulates inland and is often a nuisance because of its
erosive action and the way in which it covers highways and buildings near the
coast. Blown sand also destroys vegetation. Marram grass (*Ammophila arenaria*)
grows well in the unstable environment of sand dunes. It sends down deep roots
to tap moisture and these have a binding effect on the sand. This plant is widely
used as a stabilizer of moving sand on coasts.

Clay-with-flints

This is a residual deposit (Fig. 15) which forms above the Chalk in
England. The Chalk is a Cretaceous limestone which contains insoluble flint and
chert and a small amount of other impurities like clay and iron minerals. As on
other limestone terrain the rainfall quickly runs below the surface and dissolves
calcite, leaving behind a brownish clay mixed with the flint. These superficial
deposits are usually found as pockets in hollows on the tops of hills, or filling
solution pipes. These pipes are the underground drainage channels by which rain

Fig. 14. Desert landforms. (*a*) Barchans. Arrow shows direction of prevailing wind. (*b*) Delta fan of gravel spread by flash floods from desert uplands. Sedimentary rock at base of mountain front is capped with dark basalt sheet.

(*a*)

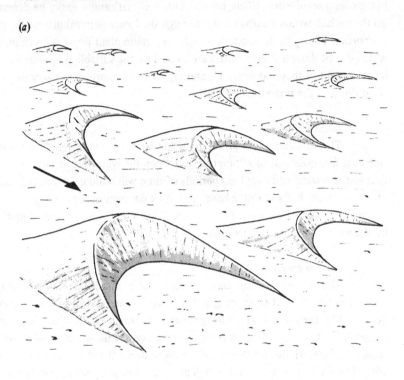

(*b*)

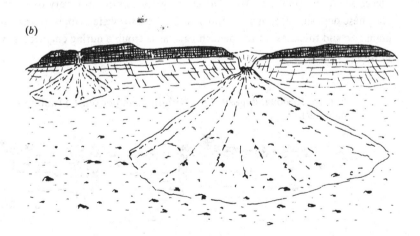

leaves the surface and are filled with an unconsolidated mass of this material. They range from 3 to 10 m in diameter and may go down as much as 20 m. They can cause obvious difficulties to the civil engineer, particularly when a highway is being constructed across Chalk terrain. This material cannot easily be detected on the surface without excavation, although the lower permeability when the proportion of clay in the mixture is high may cause plant population changes which can be detected. Difficulties are caused by the variable thickness and lateral extent of these pockets of material, which have properties very different from those of the surrounding Chalk.

Loess

The dust carried by prevailing winds blowing away from deserts carries fine dust for thousands of kilometres, but when the wind velocity drops the dust falls to the ground and over long periods of time will build up deposits hundreds of metres thick. Parts of China have very thick deposits of loess which have originated from Central Asia. The lower levels in the deposit become consolidated by the weight of the material above.

Plateau gravel

This is found in large areas, of the order of hundreds of square kilometres, and consists of layers of gravel that have accumulated on level ground, deposited by the water from melting glaciers. Another way in which large areas of gravel can form is as delta fans which spread out at the bottoms of mountain ranges in places where fast-flowing rivers come down the mountain front (Fig. 14*b*). The velocity of the water is high and only coarse gravelly rock detritus can remain in place; all the fine material is moved further away, and accumulates in places where the gradient is less. The coarse grain size makes it very permeable and these deposits often make important aquifers for water supply. Variations in grain size and thickness of the deposit can cause trouble during engineering works in these deposits.

Fig. 15. Residual deposit. Clay-with-flints in pockets (*A*) and in solution pipes (*B*).

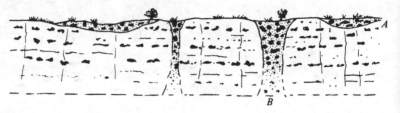

Peat

This is an accumulation of plants which have not completely decomposed. Peat forms in cold and wet conditions in which the normal recycling of plant tissue to carbon dioxide and water is retarded because oxidation cannot occur. Peat is typically found in wet ground where sub-surface drainage is poor. Water makes up a large proportion of the mass and the deposit has almost no bearing strength. There may be sudden changes from hard rock surfaces to peat deposits, the contacts are often steep and irregular because peat forms in small pockets in the ground as well as in large waterlogged areas. Problems arise when highways are made across ground that is partly covered with peat. Depth to rockhead must be found, and if this is too deep for excavation down to solid rock, pile foundations have to be used.

Solifluction deposits, mudflows, soil creep

These are deposits of weathered rock which has moved down slopes in the ground as the result of being saturated with water and becoming unstable. On the side of a hill the process can build up a series of little terraces called terracettes.

When the mass of material is very large and saturated, the movement and the deposit is called a mudflow. This type of deposit is considered in Chapter 4 in the section dealing with unstable ground.

Fill, made-ground

This is man-made, and is usually tipped material consisting of urban refuse and demolition rubble filling old quarries and other excavations and hollows in the ground. It is used to level off the surface. The waste from towns over centuries, and the practice of making up sites to specified levels, means that towns grow upwards. Urban redevelopment work will require accurate survey of this superficial material. Boundaries between fill and bedrock can be very steep to vertical, and the material can be very poorly consolidated.

Coastal deposits, longshore drift, spits, sand bars

These deposits are formed from gravel and sand produced by marine erosion of the coast, or brought to the coast by rivers. Where there is a prevailing wind which causes the waves in the sea to follow the coastline approximately, and where the tidal advance is also in this direction, the combined forces keep the material moving in one direction while it is being increased in volume by coastal erosion. Wherever sea currents have a generally uniform direction of movement and not a random direction, the sand and gravel of beaches tend to form spits across the mouths of rivers, sand bars offshore, and larger areas of sand built out towards the sea, forcing it to retreat locally (Fig. 16). The top surface of these deposits is usually only a few metres thick and above sea level, and in humid climates is colonized by plants that are adapted to living in the exposed conditions of coastal environments. The troubles caused to civil engineers by longshore drift include the blocking of river mouths, which in turn causes flooding upstream, and changes in equilibrium so that a mass of sand that has been stable for centuries perhaps starts to be eroded. The material is in a continuous process of movement and, as a result of the combination of storm force winds coinciding with a high tide, a great mass of sand can be moved in a few hours and the beach and coast profile can be very greatly altered. After the extreme conditions have ceased, the natural forces will tend to bring about a return to the previous state of sand distribution.

Coastal engineering works may upset the natural balance of longshore drift processes and cause unexpected effects which can prove very expensive to rectify. The use of groynes (Fig. 16b) to stop longshore drift from blocking rivers often causes trouble because it cuts off the supply of beach material to the adjacent area of coast. The beach deposits are moved on by sea currents, leaving a sector of the coast without the protection that a pile of gravel provides (Fig. 16c). Rapid erosion of the cliff or coastline follows. Very careful consideration should be given to the possible side-effects that any coastal engineering works may cause, when they are intended to prevent some existing trouble, or when they are part of some new development such as docks.

Conclusion

All the chemical, mineralogical and rock texture or structure factors described above for all classes of rock determine the mechanical behaviour of the large masses of rock in the ground. Geologists have a very large number of technical words to describe all the properties of rock in detail; the civil engineer finds the terminology difficult to understand and in consequence tends to avoid studying geology. He should however always remember that the mechanical properties of rock are the important ones so far as he is concerned. Rock properties should be studied with the intention of understanding the causes of the mechanical behaviour of rocks. This is not a random process but is dependent on the chemistry, mineralogy, and structure of the rock. Some rock types tend to cause more trouble on construction sites than others. The fact that a rock is hard, for example limestone, does not guarantee that it will not be a cause of trouble.

Fig. 16. Coastal deposits: (*a*) coastal landforms, spit, sand bar, and foreland; (*b*) groynes built to stop longshore drift; (*c*) beach profile. Berms built up by wave forces.

(*a*)

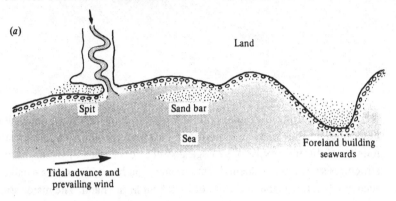

(*b*)

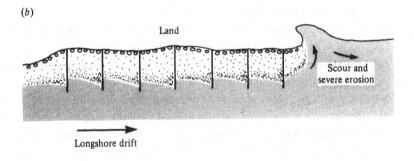

(*c*)

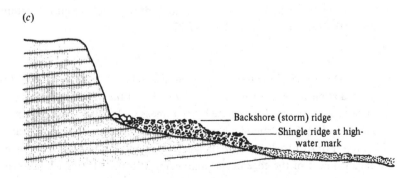

4 Geological structures, rock instability and slope movement

The term geological structure is used to cover the three-dimensional description of rock masses. The crust of the Earth is made up of a very large number of different types of geological structures. These range from very large structures such as a continental plate or an ocean basin, down to a small element such as a bedding plane in a piece of rock held in the hand. The major structures are made up of an assemblage of smaller structures, all of which have been created by the processes of sedimentation, magma intrusion, continental drift, and the rise and fall in the level of the surface of the Earth in different places. Structures may have sharp boundaries between each other, or they may slowly merge from one type to another.

The fundamental unit structure of a sedimentary rock is called the bed, or stratum. It is a layer of a particular type of rock, and although the term is applied strictly to sedimentary rocks only, it is sometimes used for any type of rock which has a layered structure. Sediments form as generally horizontal layers on the sea bed; a group of such beds is called a formation and it often consists of a number of different rock types, shale, sandstone and limestone. The word series is also used to designate a group of related rocks. After millions of years of sedimentation upward movement of the sea bed raises the rocks to form land, and this movement causes the rock formations to be tilted out of the horizontal as a dome structure forms (Fig. 17c). The beds are said to dip, that is to slope downwards from the horizontal. In a dome the beds dip radially outwards from the crest. Compression of the Earth's crust causes the beds to be folded into structures called anticlines and synclines (Fig. 17e, f).

Dip

Dip is one of the most fundamental of all geological structures; it is measured as an angle relative to the horizontal, and the direction of this slope relative to north must also be stated. Dip is measured with a clinometer placed on the outcrop bedding plane of a hard rock, and a compass is used to measure the direction. For soft rocks with no hard surfaces on which to place a clinometer, visual alignment of the edge of the clinometer with the bedding planes seen from

a distance is the method used, or levels can be taken with a levelling instrument
along a traverse of the outcrop and the average slope calculated. An example of
dip would be 40° N 045°, which means that the beds slope at an angle of 40°
below the horizontal in a north-easterly direction. A line at right angles to the
maximum slope on the rock bed is called the strike. It is a horizontal line on the
surface of the rock (Fig. 18a). Strike is also the direction of the edges of the

Fig. 17. Geological fold structures: (a) basin of deposition; (b) section
across basin; (c) dome; (d) section across dome; (e) anticline; (f) syncline.

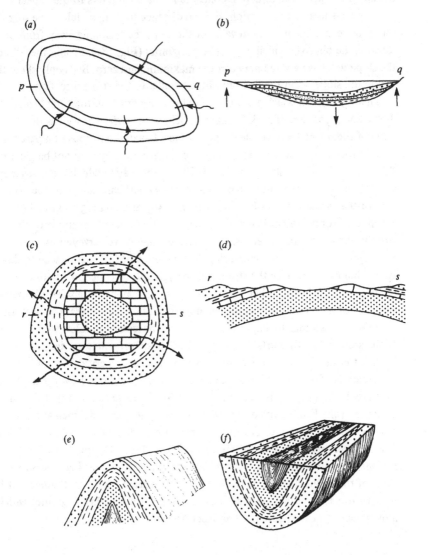

beds, and it may be compared with the grain of wood, which is also built up in a layered structure like sedimentary rocks. Geologists speak of the grain of the rock formations in the same way as a carpenter speaks of wood grain. The two concepts are so closely related that the best way of understanding the structures of folded rocks is to study blocks of wood cut in various directions to the grain. Dip is defined as the maximum angle of slope on the rock bed, but there are smaller slope values in other directions and the slope decreases as the direction considered moves round towards the strike direction, along which by definition the dip is of course zero (Fig. 18c). Any of these smaller angles of dip is called the apparent dip in the direction concerned. The term refers to the appearance of dipping beds in a vertical section. From a distance it is impossible to be sure that the slope seen is the maximum value of the slope; beds sloping away from or towards the observer will appear to be horizontal (Fig. 18a). From other directions the slope will have a value between the maximum and zero. Beds only show their maximum slope when the line of sight is parallel to the strike direction.

The object of measuring dip is to obtain information on the total three-dimensional position of rock formations below the surface as well as those parts of them which are visible on the surface. Isolated outcrops of a particular type of rock may be connected below the surface, or they may not be parts of the same bed. Fig. 19 shows one possibility of the way in which outcrops may be connected underground as part of a combined anticline and syncline. The arch of the anticline sector has been eroded away and no longer exists, but the syncline sector is present below the surface. This is a simple example of the structural analysis which geologists make after they have surveyed the area, identifying rock types and measuring dips. Where a bed of rock is seen to disappear below the surface the structural analysis will be concerned with estimating the depth of the bed at other places. The analysis may be for a scientific purpose only, or for an economic one, for example when the bed is a seam of coal. Before it could be economically exploited it would be necessary to calculate the depth of the seam below the surface at other places to determine its underground extent. Depth of a mineral deposit below the surface is a critical factor in the cost analysis of a projected mining operation. Fig. 19c shows how the depths are calculated when the dip has been measured. It of course assumes that the structure is planar; if the bed is curved the dip varies over the area concerned and the calculations will not be accurate. Tests of accuracy are made by drilling boreholes in different places and finding the different depths to the coal seam. There are several reasons why the seam may not be met at the expected depth; Fig. 19d shows three of them: the dip may have changed; or the seam may have thinned out to zero; or it may have been displaced by faulting, fracturing in the ground which moves beds in various directions as described below.

Fig. 18. Dip. (*a*) Dip and strike. As seen from the front of the outcrop
the beds appear to be horizontal, but the true angle of dip is seen when
viewed from *x*. (*b*) Clinometer measures dip as 30°. (*c*) Calculation of
apparent dip. (*d*) Dip of beds appears to decrease as viewpoint moves
round from *x* through *y* to *z*.

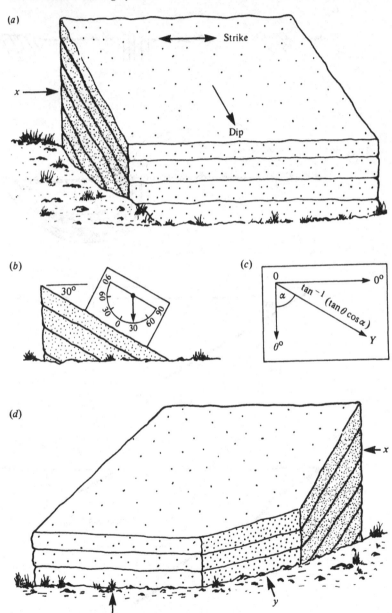

Fig. 19. Dip. (*a*) Outcrops. (*b*) Possible structure: anticline (left) changing to syncline (right) with vertical central limb. (*c*) Calculating depth of bed below surface. (*d*) Three reasons for bed not being met at expected depth.

(*a*)

(*b*)

(*c*)

(*d*)

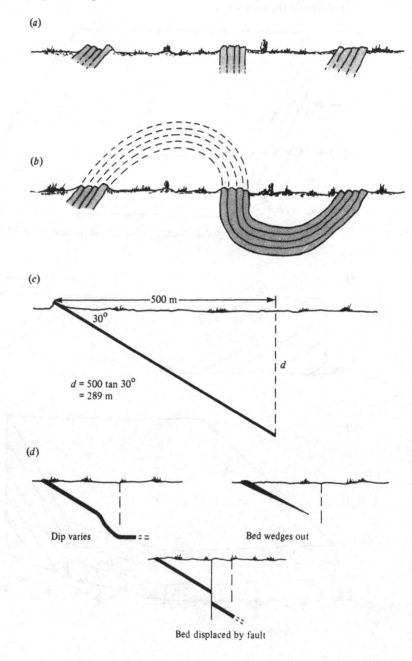

$d = 500 \tan 30^{\circ}$
$= 289$ m

Dip varies   Bed wedges out

Bed displaced by fault

It is usually found in practice that the measured dip values are not constant over the outcrop because the beds are not perfect planes but have irregular, curving surfaces. An average value can be calculated from a number of readings. This is not a simple calculation of average of the dip angles because it has to include the azimuth direction of the dip in each case. Details of these calculations are included in textbooks which describe structural geology.

The true dip can also be calculated from readings of apparent dip measured in different directions. A minimum of two readings will give the solution, but it is better to take several readings on vertical face exposures like natural cliffs, or excavations like cuttings for highways, or from borehole results. There are several methods of calculating dip from readings of apparent dip. The graphical cotangent method is recommended to the student because it helps to develop a three-dimensional sense of rock masses below ground level and on the surface. When these calculations have to be made regularly it is easier to set up a simple program to use with a calculator, based on equations connecting the trigonometry of the dip values. Fig. 20 explains the cotangent method, which is useful because it immediately shows up any anomalous or inaccurate readings, which would not be detected in the methods based on equations.

Folds

Fold structures are caused by compression within the Earth's crust related to lateral movement of continents. The strata in the zone of compression become folded into corrugated structures (Fig. 21). There is a series of folding intensity from shallow folds to very intense folding when the rock formations will be strained beyond the elastic limit and break. This process is called faulting and when under compression the rocks are sheared and overlap occurs. When all the fold axes are almost parallel to each other the folding is called isoclinal. Folding is one of the causes of discontinuities in rocks. A plastic material like clay can adjust itself to the distortions caused during the folding, and will expand into the crests of folds and thin as a result of squeezing on the limbs (flanks). The fundamental cause of the distortion is the progressively greater length of arc outwards from the centre of the fold. If the volume of material is to remain constant the bed must become thinner, or more material may move upwards from the flanks. Ultimately, if intense compressive forces act over millions of years the clay will be changed to slate (Fig. 11g). Hard rocks like limestone or sandstone behave differently. Because they are rigid bodies they must fracture, and a very large number of discontinuities are created, similar to those shown in Fig. 3. All the widths across the discontinuities when added up will be approximately equal to the extra length of arc. Fig. 21c shows the details of these joint patterns.

Analysis of rock under compression shows that it is subjected to a stress field which can be represented by three components: the active lateral force which is causing the distortion; a sideways restraining force due to the rigidity of the Earth's crust at the place concerned; and the vertically downward force of gravity. The last also includes the weight of the rock all the way up to the surface of the Earth. When the active lateral force exceeds the other two, the whole mass of the rock is forced upwards into folds, and it will also expand sideways (Fig. 21*d*). This sideways movement causes joints to form which are perpendicular to the fold axis; this is the reason why one often sees smooth rock surfaces cutting across the axes of folds, making them appear exactly as they are

Fig. 20. Cotangent method of dip calculation.

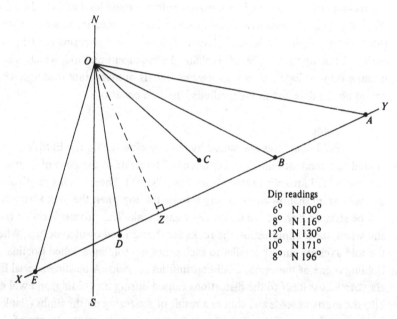

Dip readings

6°	N 100°
8°	N 116°
12°	N 130°
10°	N 171°
8°	N 196°

Method:
Draw OA on bearing 100° of length proportional to 9.5 (cot 6°)

OB	116°	7.1 (cot 8°)
OC	130°	4.7 (cot 12°)
OD	171°	5.7 (cot 10°)
OE	196°	7.1 (cot 8°)

Draw best fitting line *XY* through *A B C D E*
C appears to be anomalous or erroneous reading so neglect it
Draw *OZ* perpendicular to *XY*

$\cot^{-1}(OZ) = \cot^{-1} 5.4 = 10° 30'$

Bearing of *OZ* = N 156°

Answer: Dip 10° 30' N 156°

Fig. 21. Folding: (*a*) progressive intensity of folding; (*b*) structural parts of folds; (*c*) joint systems in a fold; (*d*) forces operating during folding.

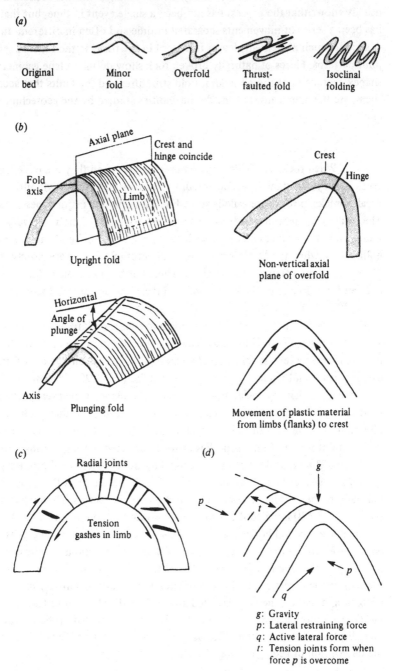

(*a*)

Original bed   Minor fold   Overfold   Thrust-faulted fold   Isoclinal folding

(*b*)

Axial plane   Crest and hinge coincide

Fold axis   Limb   Crest   Hinge

Upright fold   Non-vertical axial plane of overfold

Horizontal   Angle of plunge

Axis   Plunging fold   Movement of plastic material from limbs (flanks) to crest

(*c*)   (*d*)

Radial joints

Tension gashes in limb

g
p
t
q
p

g: Gravity
p: Lateral restraining force
q: Active lateral force
t: Tension joints form when force *p* is overcome

shown in drawings. An analysis of the joint and fold patterns of rocks enables structural geologists to find the history of the folding of the rocks in an area. This usually shows that the process has not been a single event in time, but that there has been a series of movements separated in time and often in different directions. A series of such movements leaves its record in the rocks in the form of sets of joint patterns. Pieces of naturally broken rock show definite plane geometrical shapes, rectangles or rhombohedra. Fold structures and the faults that accompany them, are the main causes of the discontinuities studied by the geotechnical engineer.

Faults

When rocks are folded by compression, or when they are pulled apart by tension, they can stand a certain amount of distortion but they will ultimately break. Breaks in rocks are called faults. Fig. 22 shows the major types of faults. The total movement in a fault can amount to thousands of metres in very large faults, and at the other end of the scale the movement can be of the order of millimetres, when the break is best called a microfault. Faults are closely related to joints and are often parallel to them. There has been a break and movement in a fault, but in a joint there has been no movement across the discontinuity plane (Fig. 22*f*).

Fig. 21*a* shows how increasing compression causes rock to break, causing a thrust or reverse fault (Fig. 22*b*). The angle (θ) of dip of the plane of the fault is quite low for a reverse fault, usually less than 45°. Tension in a rock formation can also cause faulting and produces a normal fault (Fig. 22*a*). Normal faults usually have a high dip angle, more than 45°. The plane of the break, called the fault plane, sometimes has grooves cut into it (Fig. 22*c*), and these indicate the direction of movement of the two rock masses. This grooving is called slicken-siding, and it is one of the properties of rock described for engineering purposes (table 10). The fault plane can be described by three-dimensional geometry and the total movement (slip) can be considered as the resultant of three separate mutually-perpendicular vectors. Fig. 22*c* shows these vectors. The vertical movement (vector **AB**) is called the throw, the forward horizontal vector (**BC**) the heave, and the lateral horizontal vector (**CA'**) the shift. The general case is shown, with movement components along each of the three directions. Faults can also have single components of movement, a vertical fault has no shift or heave and the fault plane is vertical. If the movement has been almost entirely in a single horizontal direction the fault is called a wrench fault (also tear or transcurrent fault). All faults cause displacement of beds and are detected in the ground when such displacements are found. This aspect of faulting is described in detail in the description of geological maps in Chapter 5.

Fig. 22. Faults: (*a*) vertical section through normal fault; (*b*) vertical
section through reverse fault; (*c*) vector components of fault movement;
(*d*) vertical section through fault; (*e*) wrench fault causes displacement
of strata; (*f*) minor faults and joints parallel to major fault; (*g*) fault
breccia in fault zone.

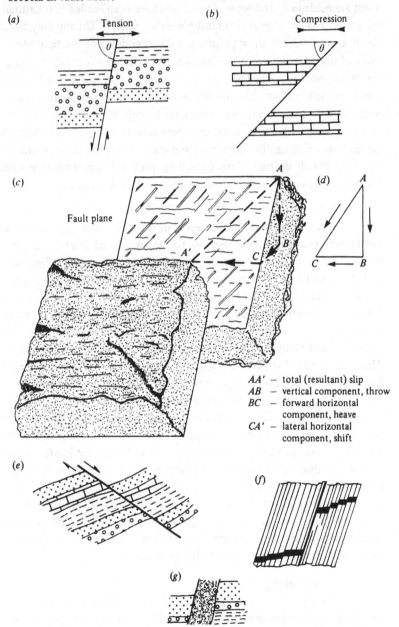

AA' – total (resultant) slip
AB – vertical component, throw
BC – forward horizontal
    component, heave
CA' – lateral horizontal
    component, shift

When the rock close to a fault is studied closely it often shows a great number of small faults with similar orientation, but with throws of only a few metres, or centimetres. These parallel trending structures, and often joints as well (Fig. 22*f*), have important effects on the stability of the rock when excavations are made in it, and when steep slopes have been eroded in it naturally. The diagrams of faults suggest that fault planes are smooth. Usually only the minor faults are smooth and show polishing and slickensiding on the fault plane; the zone of the main movement is often a mass of broken rock, called fault gouge, which can be as much as 100 m wide. This mass of broken rock in the fault zone weathers easily and usually contains many voids, filled with water during wet weather, with groundwater when the zone is deep. Water in these fault zones can flow very quickly because of the high permeability of the broken rock, and has caused many mining disasters when tunnelling has broken through into a fault. A pilot drill hole in front of the tunnelling work should give advance warning of danger of flooding. Faults can also cause rock falls in tunnels.

Joints

Joints are planes of weakness in hard rocks, and they are also found in highly compressed soft rocks like clay (overconsolidated clay). These planes usually occur in regular geometrical patterns (Fig. 22*f*), so that when pieces of rock break from the rock face they have quite distinct shapes, rectangular, triangular section prisms, rhombohedral and pyramidal shapes are common. The difference between a joint and a fault is that there has not been a movement along the plane of weakness in a joint. When rocks are stressed joint patterns are generated, and eventually movement occurs along one of them, forming a fault. This is the result of the concentration of stresses along one plane, which then becomes the fault plane. The process in rock is similar to what happens in metals which are stressed beyond the yield point.

These stresses in rocks operate over millions of years and if there is expansion of the rock, cracks will form which allow water to move through the rock, transporting and depositing minerals from solution and lining the joints. Quartz is a common mineral filling joints in sandstones; calcite is common filling joints in limestones. Both these minerals are in the form of white veins which often have regular geometrical patterns, related to the stress fields which have been active in the past. These joint patterns enable geologists to reconstruct the past events which have strained the rock, and sometimes there is a record of several separate events with stresses operating from different directions.

Unconformity

An unconformity is a geological structure in which one set of beds lies on the upturned edges of another set (Fig. 23). There is an angular discordance

Fig. 23. Unconformities. (*a*) High-angle unconformity. Upper series
starts with basal conglomerate (*w*), followed by grit (*x*), and fine
sandstone (*y*). Lower series is a sedimentary sequence of shale, grit,
sandstone, and limestone beds (*e–n*) which have been tilted, eroded,
and covered by the upper series. Basal conglomerates often contain
pebbles of the rocks of the series which they cover. *A–B* is the erosion
surface. (*b*) Geological map of an unconformity. Arrows show the dip
directions of the two rock formations. (*c*) Low-angle unconformity,
seen in coastal cliffs. At *P* the bed (*t*) of the upper series lies on
a different bed (*k*) than it does at *Q*, where it lies on bed (*l*); at *R* it
lies on (*m*).

(*a*)

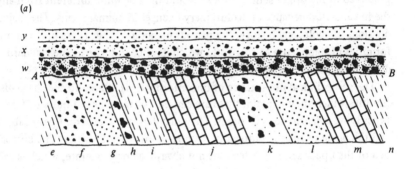

(*b*)

(*c*)

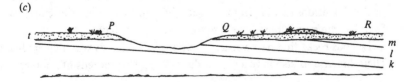

between the two rock formations. The history of this structure is sedimentation of beds *e* to *n*, followed by uplift of the sea bed, erosion to surface *A-B*, and sinking of the area again to below sea level, followed by more sedimentation which forms the upper group of rocks (*w*, *x*, *y*). When letters are used to denote rock beds it is usual to make a break in the lettering if two groups of rocks separated by an unconformity are present in the area being described. An unconformity is important in geological science because it marks an important break in geological history of the area concerned. There have been two separate phases of sedimentation, separated possibly by tens or even hundreds of millions of years. Any fossils preserved in the upper series of rocks would then be quite different from any in the lower series, because of evolutionary changes in animal forms. The major divisions of geological history, the periods named in Table 1, are based on unconformities in some parts of the world which have been taken as the standard divisions of geological time.

The geotechnical importance of unconformities is related to the sharp discontinuities which they contain; these are caused by two rocks of quite different types lying adjacent to each other. They may have very different permeabilities and the plane of the unconformity may be saturated with water. The bottom bed of the upper series is often, but not always, a conglomerate, which is called a basal conglomerate. The reason for this is that when a new land surface has been raised the sea attacks the edges and erodes its way inland. The first deposits formed will be along the initial beach: pebbles, as described in the section about beach deposits. As the sea moves further inland the first beach deposits of pebbles are located further out to sea, in an area that is receiving finer-grained deposits such as sand or silt. There is often a well-marked gradation from coarse- to fine-grained material from an unconformity upwards. On the ground an unconformity is marked by intersection of boundaries between different rock types. The pattern is similar to that seen in a vertical section of rocks containing an unconformity (Fig. 23*b*).

Igneous rock structures

Igneous intrusions make up another group of rock structures which range in size from thousands of kilometres to centimetres, from batholiths to small veins in other rocks. The very large masses are called batholiths or plutons (Fig. 5), formed in the cores of mountain ranges and now visible after erosion. This structure has been described in the section on granite in Chapter 3. The smaller intrusions coming from the batholith are called bosses or cupolas (Fig. 5*b*); they are extensions above the main mass of the batholith.

Igneous rocks that have vertical positions are called dykes; they are like walls. Their dimensions also have a very wide range up to hundreds of kilometres in length, and tens of kilometres in width, formed by intrusion of magma into large cracks in the Earth's crust. Sills are igneous structures which are dominantly horizontal, but there is no defined angle at which a dyke becomes a sill. The change point can be considered as an angle of 45° to the horizontal. Details of sills and dykes are shown in Fig. 7 and described in the section on gabbro, dolerite, and basalt. Basalt is also found in very wide sheets, covering hundreds of thousands of square kilometres. The Deccan Plateau in India is one example. The basalt is believed to have come from a major fissure in the crust. The significance of these structures to the geotechnical engineer arises from the possibility of sudden changes in rock properties. Igneous rocks are often very hard, and strong because they do not have bedding planes like sedimentary rocks which can often be split easily along these planes of weakness and along regular joint surfaces. Dolerite usually has irregular joint patterns (Fig. 8); some masses of granite on the other hand contain a layered structure, dilatational jointing near the surface. Sills and dykes of all kinds of igneous rocks are useful as sources of aggregate.

The geometrical patterns of the large structures like bedding and jointing are generally continued down the scale to microstructures within a piece of rock taken for laboratory testing. These may be visible when the rock is examined as a transparent thin section (30 μm thickness) in transmitted light under a microscope designed for geological applications. The microstructure controls the strength of core samples, which are tested under compression. The strength may be equal in all directions (isotropic), or because of mineral alignment, or microscopic changes in grain size, it may be different along three mutually-perpendicular axes, when it is called anisotropic. Slate is a good example of a rock that has anisotropic strength properties.

The macrostructures usually determine the strength of the rock mass. Bedding planes and jointing, and pockets of highly-weathered rotted rock are the controlling factors in the strength of rock on and near the surface.

Landforms

Landforms are defined as distinctly-shaped units of the Earth's surface, such as hills and valleys, cliffs, estuaries. An assemblage of landforms is called a landscape. There is a close relationship between the shape of landforms, rock structure, and the weathering properties of the rock below the surface. Climate is also a very important factor in determining the type of landform that will appear on a given set of rock type and structure conditions. During the last ten years geotechnical engineers have made increasing use of landform measurement in the assessment of the geotechnical properties of a site which is being investigated before construction starts. They are particularly concerned with the stability of the ground below construction sites, and with the prevention or control of landslides, as examples. Large, linear constructions, highways for example, require early assessment of the terrain before the final choice of the route. The applications of morphological mapping, the measurement of landforms, will be described under the heading of geotechnical mapping. The science of quantitative measurement of landforms is called geomorphology.

Types of landforms have been briefly described in connection with weathering processes, erosion, and in Chapter 3 in the description of rock types. Some rock types give rise to very distinctive landforms, for example limestone weathers to form karst.

The geotechnical engineer with a good understanding of the weathering and erosion processes related to climate and rock type will be able to see important details in the landscape, particularly those which are indicators of rock instability, traces of slow movement of soil which would apppear to be stable to the untrained observer. There are many clues which give advance warning of future landslides. The underlying rock may be completely hidden by the cover of soil produced by weathering and drilling may be necessary to determine the rock type below. In arid and humid climates the strength of the rock controls the angle of the surface slope and hence the shape of the ground surface. Hard rocks like limestone and sandstone stand out as convex curves in contrast to the concave curves formed on soft rocks like clay or shale. Relative strength and the resistance to weathering and erosion are the important factors. In humid climates the profile consists of smooth curves which change from concave to convex through points of inflexion (Fig. 24a), but in arid climates the profile is usually much sharper, with angular intersections between different slope angles (Fig. 24b). The landforms of glaciated regions also show sharp angular changes between different slope angles, but these are more dependent on where the ice has happened to accumulate than on rock type because ice can easily erode hard rocks, especially when they have been broken into small pieces by the expansion force of water when it freezes. When the ice has melted and disappeared as the climate becomes warmer and humid, the angular profiles tend to become rounded.

The slope angles that are formed on soft, unconsolidated rocks are dependent on the grain size of the material, its permeability, and average moisture content during the whole year. Although this may change with the seasons, it is the average that matters. For each set of parameters there is a maximum slope angle that is stable. For a given grain size and permeability the variable with respect to time is the water content of the rock. When the ground is saturated it has a lower angle of repose (Fig. 24c) than when it is dry, but the exact relation between the variable factors is complex and surface tension between liquid and solid is critical.

Any particular landform is undergoing a continuous process of change, even though it may be an extremely slow change. The whole period of history of erosion of the land can be described as a sequence of youth, maturity and old age. A youthful landscape is one in which the agents of erosion have been acting for a relatively short time on a recently uplifted land mass. The starting surface may be a former sea bed or a volcanic terrain. This type of landscape has wide flat surfaces cut into by rivers which have narrow valleys with steep sides (Fig. 25*a*). Because of the steep slopes the erosion is chiefly downwards. After millions of years the valleys widen and meet adjacent ones, so that only small parts of the original flat uplifted land area remain. This is called the mature stage in the erosion cycle. A landscape in old age is almost flat again, with isolated hills standing up above a plane which is almost at sea level; this is called a peneplain by geomorphologists (Fig. 25*a*). Sea erosion tends to cut down the whole land surface to sea level. The erosive power of rivers is greatly reduced

Fig. 24. Erosion profiles and rock type. (*a*) Erosion profile of gently dipping hard and soft rocks. Humid climate. (*b*) Typical arid climate erosion profile. (*c*) Angle of repose of sand.

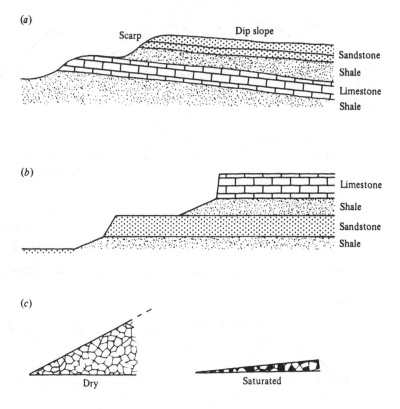

when their gradient is low in the old age stage, and their work is then almost entirely restricted to removing fine-grained material produced by weathering. All land masses would ultimately be reduced to sea level and disappear if this process continued indefinitely, and their rock material would be spread over the sea bed as silt and clay, or gone into solution in the sea water. These processes do not have a linear rate with respect to time, but are governed by exponential functions of time, and in the final stage the erosion rate is so slow that crustal movements intervene before the asymptote (sea level) is reached: the land either sinks below the sea, or it is uplifted again (rejuvenated) and the cycle starts again on a nearly flat surface related to a former sea level, but now possibly 100 m higher. When the levels of a landscape are carefully studied, wide flat areas are often seen quite clearly (Fig. 25b). These former erosion surfaces may not be continuous on the ground but may be isolated from each other. This matching

Fig. 25. Landscape profiles. (*a*) Stages in the erosion of a land surface. Vertical scale exaggerated. (*b*) Former base levels seen in a landscape profile. (*c*) Summit accordance.

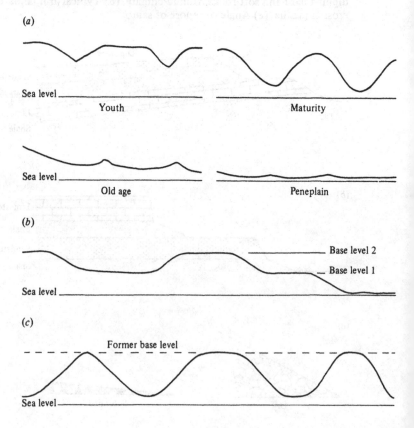

of the tops of hills and mountains is called summit accordance (Fig. 25c).

It seems that the rejuvenating process of uplift does not proceed very gradually, when erosion would keep pace with it, but there is evidence that suggests there are relatively rapid and isolated periods of uplift separated by long periods of stability, called stillstand. Some areas of the world show quite rapid movements, others are much more stable. Mountain areas are active; very large, flat areas within the continents are stable, but even these may show some movement. During the last ice age (Pleistocene) northern Europe was covered by a great mass of ice, possibly thousands of metres thick in places, as in Greenland today. The weight of this ice depressed the land into the crust slightly, and since the ice melted there has been a gradual upward movement recorded on gauges at tidal measurement stations. The Baltic Sea in northern Europe has been receding from the coast and this progress is recorded in the positions of former boat moorings a long way inland from the present coastline.

Sea levels are of course affected by ice ages because of the amount of water removed from the oceans to form icecaps. Calculations have shown that if all the permanent ice on land areas were to melt, world sea level would rise by about 50 m, flooding much of the inhabited land in the world.

These aspects of geomorphology are not of scientific interest alone; they have important geotechnical applications. The velocity of a river varies throughout its course from the source to estuary. The volume of water and the shape of the river channel control the velocity, and some stretches of river may have relatively high velocities where the profile is the link between two erosion levels (Fig. 12e). Diagrams of these profiles need to be greatly exaggerated in the vertical scale to make them clear. The sharp changes in slope, called knickpoints, are sufficient to make the river flow more rapidly over these stretches and erosion rates will be greater. The river is working to eliminate these knickpoints and to grade itself to a single exponential curve with an asymptote at the present sea (base) level. This is the background to any problem that a civil engineer may be working on in connection with river control to remedy erosion or flooding. The very large construction project in London, the River Thames Barrage, is intended to meet the possibility of flooding of that city because the area is very slowly sinking relative to the level of the North Sea.

Rock instability and slope movement

Unstable conditions range in size from a collapse of ground of about 5 m size under a building, or a small rock fall from a cliff or river bank, to a major landslide when a large part of a mountain falls into a valley. The causes of instability may be natural or man-made. Earthquakes can cause very large landslides in mountainous country. Man-made excavations, for example highway cuttings, are designed so that the batter (slope) will remain stable. The design is based on the state of weathering and the structure of the rock, and the engineer responsible for the design has to steer between an angle that is so low that instability if it should occur will cause no trouble, and an angle that does not require a very large land-take which will add greatly to the cost of the work.

Instability may be continuous and gradual or it may occur in isolated events, separated by 100 years or more. Man-made cuttings may last for more than 50 years showing no sign of trouble, then suddenly collapse. Natural instability in the ground is very much controlled by groundwater conditions. When the rock mass is dry the friction along the discontinuities is high and the mass is stable. If there has been abnormally high rainfall the water table rises and there is an internal disruptive force acting against the discontinuities. Should any movement start in hard rocks as a result of gravity force, sliding takes place along open discontinuities which may be filled with wet clay produced by weathering of the rock. In soft rocks there is a gradual movement of material down the slope, faster when the ground is wet, slower when it is dry. Clay when saturated can behave like a liquid and rapidly moving mudflows occur. A profile of stability is built up by these natural processes, and any cutting made in this natural slope is inherently unstable and natural forces will start working to smooth the profile so that it matches that for the rock type and climate (Fig. 26). The process may take hundreds of years, or it may take only days. Any geotechnical assessment of a site for a cutting should include measurements of natural slopes in the area and notes on how these vary with rock type. The area should also be searched for signs of ground instability using aerial photographs possibly, and accurate levelling to mak contoured plans with smaller contour intervals than may be included on existing maps of the area. Geomorphologists are being increasingly employed for the assessment of the stability of slope and the design of cuttings and foundations on ground in rock that is inherently unstable. Fig. 27a shows a plan of contours at 1 m intervals which show bulges of the land surface downhill into the valley. Such convex areas are very characteristic of slow creep of the material down the slope. They may be caused by other factors, change of rock type for example, but this type of profile should warn the geotechnical engineer of the possibility of unstable ground. The instability is usually accompanied by a higher than average

Fig. 26. History of an excavation, assuming no maintenance over a period of about 200 years.

Original excavation   Rockfall   Final profile

water content in the ground, and this is often indicated by a change in the types
of plants growing there, or more rapid growth of plants in the wet areas compared
with those in adjacent drier areas. Variations in the water content in the ground
are in turn often related to rock type changes or major discontinuities like faults.
Wet zones may prove troublesome during construction work and may cause
intermittent trouble to cuttings and lead to maintenance costs unless the trouble
is cured at source. Fig. 27b shows the mechanism of these minor slips in wet
ground consisting of fine-grained soft rock. Fig. 27c shows a common cause of
trouble, where a bed of permeable sand overlies less permeable or impermeable
rock like clay or mudstone. The water draining down through the sand concentrates at the boundary between the different rock types and causes the sand to

Fig. 27. Unstable ground: (a) contours; (b) small bulges in hillside
caused by slips; (c) permeable sands overlying impermeable marl.

(a)

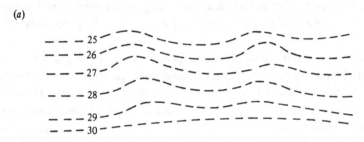

(b)

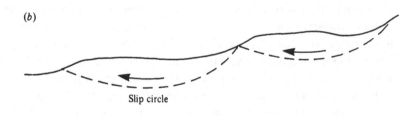

Slip circle

(c)

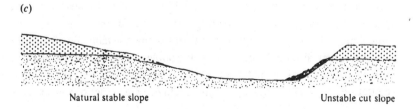

Natural stable slope   Unstable cut slope

become mobile and slowly creep downhill over the clay, a process known as spring sapping. Water running over the surface of the sand in the cutting also erodes little channels, which deepen and lead to sideways erosion and the formation of little deltas of sand running down over the cutting to the base (Fig. 10). One remedy for this trouble is to excavate small V or Y pattern drainage channels and pack them with pieces of rock about 20 cm size. Rapid subsidence all along the cutting is also possible, and depends on the possibility of the permeable sands becoming saturated during unusually wet weather.

Clay is a cohesive material and has the ability to stand up at quite high angles, sometimes vertical if dry and the height of the cutting is not so great that the load at the base is greater than the shear strength of the clay. Other clay masses are very unstable. The reasons for the difference must be looked for in the microstructure of the clay, joints in it, or in the variations in the permeability of different parts of the whole rock mass. The origin of the water causing the clay to become mobile has to be found; it may be groundwater coming from several kilometres away and flowing down a very low angle of dip in the beds making up the whole rock formation.

There is sometimes a difference of opinion on the part of geological surveyors about the rock name for this type of fine-grained material. Some beds of marl are so well indurated that they are better described as mudstones. These may be almost impermeable when in the ground and break up into large lumps of hard material when excavated, but after a short period of exposure to a humid atmosphere they will absorb water and crumble into small pieces. This will show what can be expected to happen on the surface of the cutting as a process of slow but continuous deterioration. The variations in the degree of induration of marl may be caused by the partial cementation of the clay minerals by iron compounds, or calcium sulphate. Such variations due to natural causes at the time of deposition will not usually be considered during the normal type of geological survey, the whole outcrop being mapped as marl, which it is. This is an example of how the work of the geologist differs from that of the geotechnical engineer.

Black clays have also given trouble in humid climates after 50 years or more exposure in cuttings without any instability. A possible cause of trouble here is the gradual oxidation of iron sulphide (which gives the clay its colour) to the sulphate, with consequent expansion and very gradual loosening of the clay inwards from the exposed surface, which in turn allows the surface water to penetrate further and bring in the required oxygen for the chemical process. The whole mass of clay in the cutting can then become unstable. Small climatic variations may also be the cause of rock trouble in clays and in all other rock types. The temperate climatic zones have slow changes in temperature and precipitation, sometimes several years in a sequence are colder or wetter, or both, than the average. Such unusual conditions can be a cause of rock trouble in ground that has never been known to be unstable.

The fundamental factors that control the stability of rocks, coefficient of friction, shear strength, resolution of forces into vector components and moments of forces about a point or line, are included in the textbooks of engineering science and will not be described here. The application of these principles to the problems of rock stability are described adequately in the textbooks about rock and soil mechanics quoted in the Bibliography. The basic facts are as follows. The application of static and sliding friction theory based on the well-known block of material on a sloping or horizontal surface, and the derivation of the coefficient of friction (μ) and the angle of friction (ϕ), where $\mu = \tan \phi$, to the analysis of all the forces operating within a rock mass is complicated by the nature of the material, often variable over the surface considered. The single plane surface of contact assumed in the basic equation is replaced by a very large number of separate contacts in the real material. The grain size is very important in determining the consequent behaviour. Two major categories of unconsolidated material are specified: coarse, granular soil (also called frictional soil); and fine-grained cohesive soil like clay. The strength of granular soil is built up by frictional forces between the grains where they touch each other, and this will depend on the shape of individual grains, their size distribution, and the way in which they are packed together. The degree of packing is controlled by sedimentation conditions at the time of formation, and to some extent by the pressure on the bed during its subsequent history. Clay behaves differently because of other forces, for example electrostatic and surface tension forces, which bind individual clay crystals together. The microstructure of clays is very important here. Because the material has a natural origin and has not been sorted into different grain sizes by controlled processes, there is a third possibility of engineering soil type in which both frictional and cohesive forces are operating, in fact there is a continuous gradation from one to the other. Fig. 28 shows some varieties.

The concept of a sliding block on a tilted surface is applicable to hard rocks containing bedding and joint structures. Fig. 29a shows jointed limestone dipping at about 20°. The mass is stable at increasing dip angles (θ) until it reaches (ϕ), the angle of friction, when it will slide downhill under its own weight (Fig. 29b). The coefficient of friction (μ) depends on the nature of the surface below the bed. One rock bed is separated from the next by a thin layer of some other type of rock material, otherwise it would be indistinguishable from the rock below, all forming one bed, and so on. These thin layers separating one clearly visible bed from another are called seams, and they consist of water, air, and solids which may be clay or some other unconsolidated mineral material deposited during short periods of different sedimentation conditions from those which produced the bulk of the rock. Joints crossing the bed may also be filled with a solids-water–air mixture which exerts an adhesive force across the joint plane at some times, but at others the water content may be high and repulsion forces may arise. The block will remain stable as long as the forces preventing sliding are greater than the gravitation component ($W \sin \theta$) resolved along the bedding plane, but if water enters the joints and bedding planes the frictional and adhesive forces will decrease and the block will slide. Theoretically the block is considered as a single piece, but in practice it may break across intersecting discontinuities and this makes the theoretical analysis of the conditions for limiting equilibrium more difficult. The practical application of theory to actual rock conditions where the proposed excavation is in a mass of rock of variable types, clay, limestone, sandstone, partly indurated sand transitional to hard sandstone often results in the use of a very high factor of safety, sometimes the slope advised is less than the natural slope of the ground in the area. The geologist or

Fig. 28. Texture variations in sand: (*a*) rounded sand grains, narrow particle-size distribution; (*b*) angular sand grains, narrow particle-size distribution; (*c*) rounded sand grains, wide particle-size distribution; (*d*) angular sand grains, wide particle-size distribution.

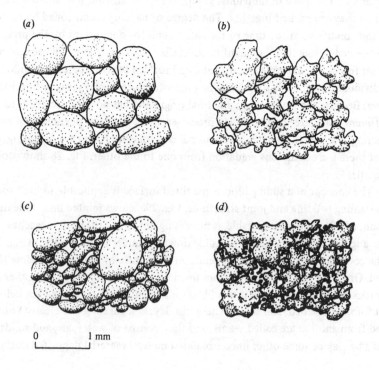

Fig. 29. Stability of jointed hard rock. (*a*) Forces acting on potentially unstable block *XY*. (*b*) Shear test. Shear stress/normal stress values of τ to cause sliding of one block over another, with increasing normal stress σ. (*c*) Graph of results obtained from (*b*). c (cohesion) is found by projecting to the axis. $\tau = c + \sigma \tan \phi$.

(*a*)

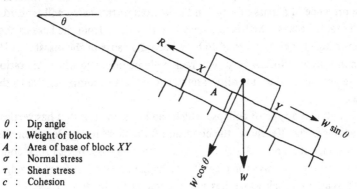

θ : Dip angle
W : Weight of block
A : Area of base of block *XY*
σ : Normal stress
τ : Shear stress
c : Cohesion

(*b*)

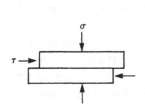

Forces acting on sloping block of weight W and base area A

Normal stress across base $= \sigma = \dfrac{W\cos\theta}{A}$

Force $W\sin\theta$ acts to cause block to slide. This is opposed by force $R = \tau A$

Shear strength of surface of contact $XY = c + \dfrac{W\cos\theta}{A}\tan\phi$

$$\tau A = cA + W\cos\theta\tan\phi$$

Block is in condition of limiting equilibrium when $W\sin\theta = R$

$$W\sin\theta = cA + W\cos\theta\tan\phi$$

If cohesion $c = 0$, $W\sin\theta = W\cos\theta\tan\phi$

$$\tan\theta = \tan\phi$$

$$\theta = \phi$$

The effect of water pressure in bedding pane *XY* is to reduce c and oppose the normal stress σ

the geotechnical engineer has to carry a very heavy responsibility in such circumstances and the factor of safety specified may lead to unnecessary cost.

Groundwater conditions, which have a controlling effect on the stability otherwise determined by the rock conditions, may change as the result of engineering works. One example is the effect on the groundwater in the hills or mountains adjacent to deep valleys which have been dammed to make reservoirs. The presence of a mass of water in the bottom of the valley will tend to raise the water table because the free drainage into the valley from the base of the hill has been reduced. This is believed to have been the cause of the disaster at Vaiont dam in Italy in 1963. A very large rock collapse occurred where dilatational joints in the bottom of the valley intersected another set of joints parallel to the valley sides.

Toppling is a type of collapse which can occur in rock that has vertical bedding or cleavage. Fig. 30*a* shows the condition for a block to topple, when the vertical line through its centre of gravity passes outside its base. In real rock conditions the disturbing sideways force for this to happen may come from a wedge of soil from weathered rock which has accumulated at the bottom of a vertical discontinuity like a bedding plane (Fig. 30*b*). This exerts a sideways force which becomes more effective as the mass of soil increases. Water in the soil helps to provide more pressure until the block topples, and the process is repeated at the next bed, which has lost its lateral support.

Rock with vertical cleavage can also fail by crumpling when the strength at the base will not sustain the weight of the rock above, after a cutting has been made. Fig. 30*c* shows the history of this type of collapse. Surface drainage water entering the cleavage planes from the top will eventually penetrate the whole mass and very gradually force the outer layers outwards, reducing the total strength. The process is cumulative until failure occurs. Efficient drainage for the surface water on the hillside and possibly sealing of the cleavage by guniting for some distance back from the cutting may be preferred as a preventive measure, instead of the removal of a large mass of rock to make a smaller slope angle (Fig. 30*d*). Rock bolting can also be used.

It is a mistake to assume that beds which dip into a hill or are horizontal will always be stable. This belief can be quite wrong. Fig. 31*a* shows beds with only a few joints, well spaced and discontinuous, which have good stability, although it is not possible to make sure of this from a superficial observation of the cut surface. Figs 31*b, c* show how such a rock structure can be potentially very unstable. There may be alignment of vertical joints, or an undetected fault in the rock within the hillside or behind the cut, which can provide a slip surface.

The way in which the rock material moves is a basis for the classification of landslides. A translational or planar slide is one in which the moving rock mass

Fig. 30. Instability in vertical rock structures: (*a*) block about to topple;
(*b*) toppling as the result of wedging action; (*c*) collapse of rock face
after excavation; (*d*) measures to prevent collapse.

(*a*)

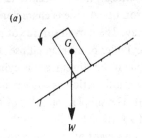

(*b*)

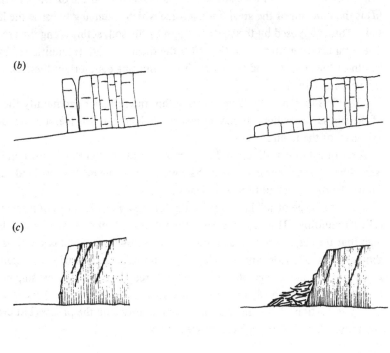

(*c*)

(*d*)

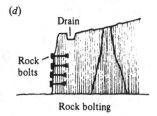

Rock bolting

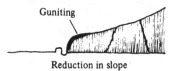

Reduction in slope

slides along an essentially plane surface (Fig. 32*a*), although it may be roughened by closely-spaced joints crossing bedding planes or cleavage. Conversely, when the failure plane is a fault it may be crossed by bedding planes. The direction of the moving mass is along a straight line (translational motion). This is the typical slide in hard rocks. A rotational slide is characteristic of potentially plastic fine-grained material like clay. The curved surface of motion is a part of a spherical surface, or possibly a cylinder or even an ellipse, so that the plan view of the slip is a curve, or straight line if the slip surface is cylindrical. Fig. 32*b* shows a typical spherical slip, seen in vertical section as an arc of a circle. The moving mass has a centre of rotation (0). The simplest analysis of this circular slide is made by taking moments about a vertical line through the centre of gravity of the moving mass. The product of the weight of the soil mass (W) and the distance to this line (d) is the moment of the gravitational force which is tending to cause the land-slide. This is opposed by the shear strength of the soil, acting along the arc (*PQ*). The total resisting strength of the soil is the shear strength (τ) multiplied by the length of the arc (*PQ*) and the radius (r). At limiting equilibrium these two moments are equal.

Fig. 32*c* shows how bedding planes within the soil mass can modify the profile of the slip surface when a harder or coarser-grained bed provides a plane surface over part of the failure.

A set of rotational slips may form a multiple regression slide, shown in Fig. 32*d*. When these occur in cliffs the backward tilt of the broken flat land surface shows the rotational motion very clearly.

The rapid type of soil landslide lasting for only a few minutes or hours is called a mudflow. This happens when the water content of a mass of clay is so high that the clay behaves as a liquid and flows downwards and laterally as shown in Fig. 33. This type of slide is often found in an area where the geological structures and the topography allow water to seep into the clay, causing permanent trouble. There is probably no satisfactory solution because of the very large mass of clay in motion. Retaining walls may temporarily stop the process but eventually they will be overtopped and pushed away.

Fig. 31. Variations in rock stability: (*a*) stable; (*b*) aligned joints: potentially very unstable; (*c*) hidden fault: potentially very unstable.

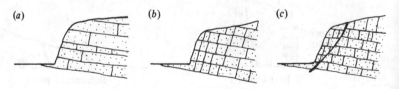

(*a*)   (*b*)   (*c*)

Fig. 32. Types of landslide: (*a*) planar translational slide; (*b*) rotational slide; (*c*) planar-rotational slide; (*d*) multiple regression slide.

(*a*)

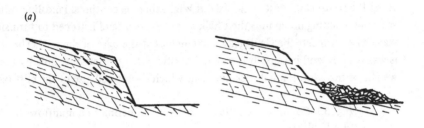

(*b*)

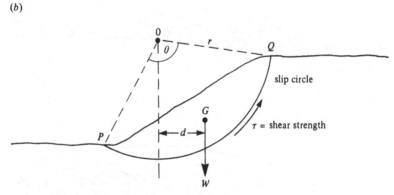

slip circle

τ = shear strength

Disturbing moment = Wd
Resisting moment = shear strength × arc PQ (= $r\theta$) × radius = $\tau r^2 \theta$
At limiting equilibrium $Wd = \tau r^2 \theta$

(*c*)

(*d*)

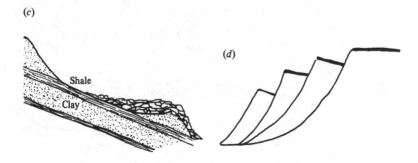

Shale
Clay

Some places have rock conditions which make them generally unsatisfactory for any kind of building work, domestic or industrial. For example, the Cretaceous rocks of south-east England contain a clay–sandstone–limestone sequence (Fig. 34), and when these beds outcrop in a cliff as at Folkestone in Kent, and along the south-east coast of the Isle of Wight there is continual landslide trouble. Water percolating down into the Chalk and the Greensand is forced to turn sideways when it meets the Gault clay and emerges at the base of the cliff. The clay is washed out, and because it is wet and plastic it is squeezed outwards by the weight of the hard, jointed rock above it, which causes collapse of the cliff face

Fig. 33. Composite mudflow. (*a*) Vertical section of a mudflow. (*b*) Plan of a mudflow.

(*a*)

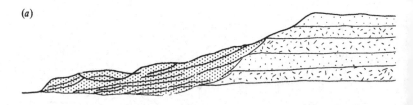

(*b*)

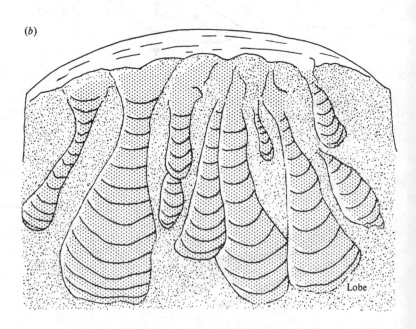

at regular intervals. There is a continual process of large blocks breaking away and sliding down on the clay mud. Sea erosion is continuously removing the debris and the cliff recedes. If there was no sea erosion at the toe of the slide the processes would eventually smooth out the sharp vertical profile to a gentle low angle slope above a mixed mass of the three rock types present in the slip zone.

The application of mathematical analysis to the conditions found in a real rock fall can be difficult.

Fig. 34. Cliff instability in Cretaceous rocks, Isle of Wight.

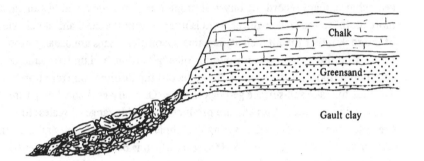

5 Geological and geotechnical maps

Geological maps

Geological maps record information about the type of rock in an area; geotechnical maps record the physical properties of the rock, and in general are more useful to the civil engineer, who is interested in the mechanical behaviour of the rock rather than its scientific name. Geological maps are usually produced for rock on the surface, but for mining, oil exploration, and underground construction purposes it is advantageous to record the information from boreholes and thus to make sub-surface geological maps. Ordinary geological maps are available in many countries and are produced in a wide range of scales; for example from 1:10 000 000 covering the whole African continent and the Middle East in a single sheet, to the 1:50 000 scale which is best for general scientific geological purposes. For constructional engineering, these published maps are often unsatisfactory because the scale is not great enough for every geological fact to be included. The civil engineer is interested in the geology of his construction site, not so much in the rocks of the surrounding area. Therefore special geotechnical surveys are made, in order to record all the details which may be of use when the project is designed, and when the work is in progress.

The surface of the ground is covered to varying amounts with a superficial cover of weathered rock, river deposits and rock debris of glacial or wind action, all of which is collectively called 'drift' by British geologists. Geological surveyors distinguish between the 'drift' deposits and the 'solid' rock formations below them. The latter may actually be unconsolidated rock material like clay or sand, but are of much greater age compared with the superficial drift, which has been produced during the last two million years approximately.

Geological surveyors start work with a topographic 1:10 000 map and record on it all the outcropping rocks, noting the different types by symbols. Outcrops in regions of weathered rocks are usually discontinuous, the space between them is covered with drift, mostly weathered rock except where there has been much deposition from melting glaciers. After a preliminary survey of the area to study the topography and to collect specimens from outcrops a decision is made about the number of different rock types to be recorded on the final map, that is a decision based on rock classification. It is not easy to separate rocks into distinct classes because each species tends to grade into the next similar one — there are far more hybrids than in plant and animal classification systems. For example pure quartz sandstone is an end member of a series which grades into calcareous sandstone when the calcite content increases, and into sandy (arenaceous) limestone, then into pure limestone when the composition is almost entirely calcite and no quartz. Other varieties of limestone contain clay, and as this increases relative to calcite the rock grades into shale or clay with impurities.

When the types of rock to be mapped have been agreed the whole area is thoroughly surveyed, all the outcrops are noted, dip of bedded rocks is measured, and all relevant information is recorded in a notebook at each outcrop. Dip is mapped by an arrow pointing in the direction of the dip (downwards), with a figure to record the dip angle. Letters or symbols are used to record the rock type on this field map according to a prearranged system. Bed thickness is measured.

The next step is to mark in the boundaries between the different rock types. A geological boundary is a surface which separates rocks of different types, on a map it appears as a line. When the place where one rock type ends and another starts can be clearly located, the boundary is marked on the map with a solid line. If the separation between the rock types is not clearly seen, often because of drift cover, an interrupted line is used to show the boundary. It may be necessary to excavate trenches or dig pits through the superficial cover if the boundaries are to be accurately recorded. The accuracy of the survey depends on the time and money available for the work. If these factors exclude the use of excavations the geologist has to rely on his experience of the way in which minor variations in the slope of the ground are related to the type of rock below — hard rocks tend to produce convex surfaces, soft rocks like shale or clay weather to concave surfaces. The boundary between them is taken as being along the line where the surface changes from convex to concave. This technique is known as feature mapping, because the hollows and raised parts of the surface are called topographic features. One example of a feature would be a small unit of landform like a 10 m wide ridge running across a hill, caused by a band of hard rock below. The boundary between one rock type and another is mapped by walking along the edge of the topographic feature believed to be related to the rock concerned, and locating this line on the field map. Figs 35, 36 and 37 show the stages in the making of a geological map.

Less accurate possibly, but much faster methods of making geological maps are based on aerial photographs and stereo-plotting. Infra-red film is also used, because different rock types have different thermal properties and hence have variations in thermal radiation. Satellite surveys can also be used to make global

Fig. 35. Geological mapping. Stage 1, outcrop information recorded on field map (topographical symbols have been omitted for clarity).

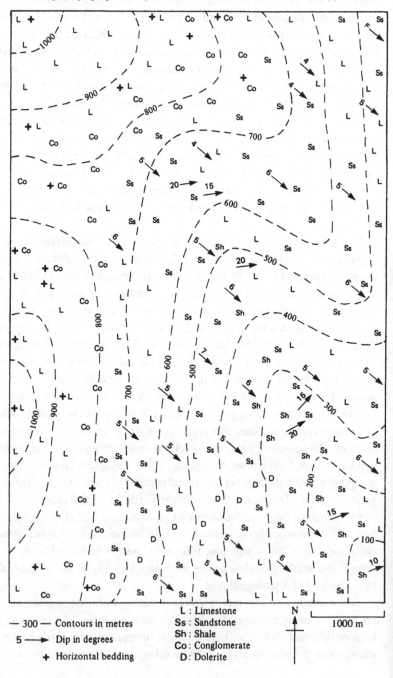

— 300 — Contours in metres

5 ➤ Dip in degrees

+ Horizontal bedding

L : Limestone
Ss : Sandstone
Sh : Shale
Co : Conglomerate
D : Dolerite

N

1000 m

Fig. 36. Geological mapping. Stage 2, boundaries inserted.

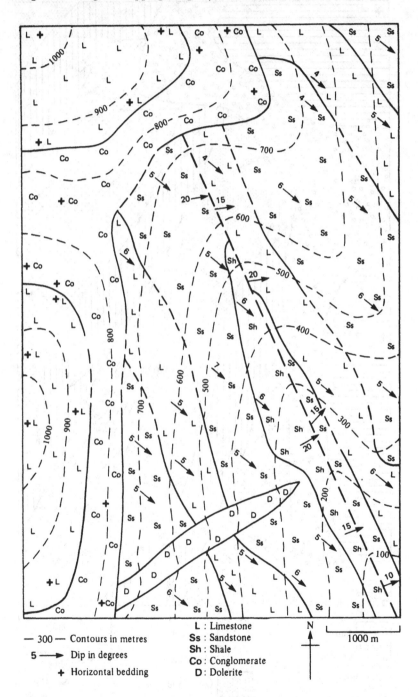

— 300 — Contours in metres
5 ⟶ Dip in degrees
+ Horizontal bedding

L : Limestone
Ss : Sandstone
Sh : Shale
Co : Conglomerate
D : Dolerite

N

1000 m

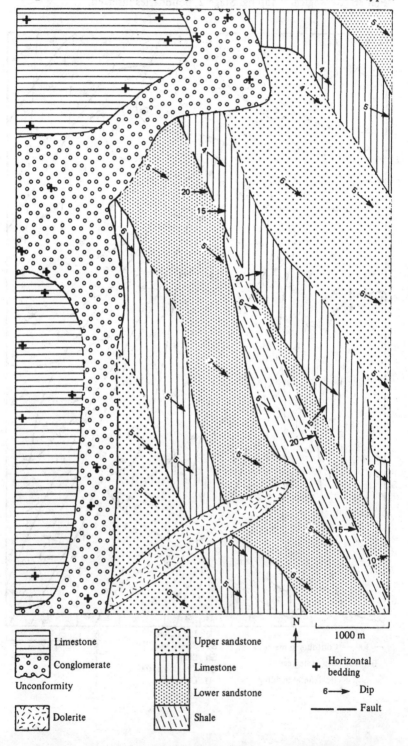

Fig. 37. Geological mapping. Stage 3, completed map. Published maps for general use are normally overprinted in colours to indicate rock types.

Limestone		Upper sandstone	
Conglomerate		Limestone	
Unconformity		Lower sandstone	
Dolerite		Shale	

N

1000 m

Horizontal bedding

6 → Dip

—— Fault

geological maps. A careful study of the types of plants, trees and smaller plants, is used because some plants are sensitive to groundwater conditions, or to trace elements present in the soil, and these properties are in turn related to the rock type. The Earth's magnetic field and gravitation force have minute variations caused by rock type and these can be detected on the very sensitive instruments designed for this work. Ground parties of geologists are sent to the area if more detail is needed, and if the purpose is to locate mineral deposits, drilling equipment will be used.

The results of the survey are published as geological maps, which are ordinary topographical maps overprinted with colours to show the different rock types. Solid lines are used to mark the position of faults, interrupted lines when the position of the fault is not accurately known, as for geological boundaries. The commonly used scale is 1:50 000. A memoir, or description, of the geology of the area included in the map is generally published as well, and this includes much descriptive material about the rocks, variations in type, fossils and minerals, and often photographs and sketches of important outcrops. When a geotechnical survey is needed the first step is to consult available geological maps and memoirs of the area. Much geological information is also recorded in the publications of the geological societies of the world.

The scales of 1:10 000 or 1:50 000 are usually too small for geotechnical purposes because they cannot show the necessary detail; roads, etc. have to be shown much wider than they actually are in order to achieve clarity. Special site surveys are therefore made, using boreholes, augering and trenches to find details about the sub-surface rock conditions. The results are recorded on plans to scales of 1:2500 or 1:1250.

When studying a published geological map the student should note the following details:

1. Title and serial number. The title is chosen from names of towns or villages in the map area; the serial number is one from a grid, usually covering the whole country concerned.

2. 'Drift' or 'solid' is noted on the map near the title.

3. Geological column and key to the colour code for the rock types.

4. Lettering system for designating rock formations on an age basis. These are standard for the geological survey organisation in the country concerned. These letters and subscript or superscript numbers are printed on the sheet; colour-blind people can therefore use geological maps.

5. Dip values are shown as arrows indicating direction and numerals for the angle in degrees. The point of the arrow marks the point of the dip observation.

6. Faults.

7. Unconformities; shown by geological boundaries meeting at an angle.

8. Igneous intrusions: plutons, sills, dykes, volcanic necks. Igneous rocks are shown in the key separately from sedimentary rocks, and are designated by letters.

9. Sections may be given to help with the interpretation of the rock structures. Lines of the sections are drawn across the map.

10. Drift deposits are grouped separately from the other rocks in the key.

11. Interpret the geological structures in the area by studying the patterns of the outcrops of the different rocks.

The civil engineer will probably only need information on the kind of rock occurring at the site, but interpretation of the geological structure of the area is often useful and interesting. The area geological structure is very important for groundwater studies. If the area contains faulted rocks, the geometry of the faults should be studied. There is often a close relationship between the directions of the faults in the area and the discontinuities in the rocks at the site. The reason for this has been stated in the section dealing with joints and faults.

The patterns of the geological boundaries on maps are determined by two factors: the geological structure and the topography of the area. The map is a projection of a set of curved and plane surfaces, and these intersect the three-dimensional mass of the rocks. The broad pattern seen on a 1:50 000 geological map for example is determined by the structure of the rocks, the folds, faults, unconformities and intrusions, but careful study of the boundaries shows that the path followed by them makes small diversions from the general direction. This second order variation in the shape is caused by the intersection of two surfaces; one is the plane of the rock boundary surface, the other is the variation in the ground, which consists of hills, ridges, and valleys. If the ground were perfectly flat the pattern would be determined by the rock structure alone, as shown in Figs 38*a*, *b*, *c* in plan views and elevation.

U-shaped outcrops indicate plunging anticlines and synclines, and to find which of these causes the pattern one applies the principle that older rocks dip under newer rocks (unless the whole structure is an overturned fold). The relative ages of the rocks are found from the geological column at the side of the map. Fig. 38 gives the relative ages as numbers, 1 being the oldest, 2 the next oldest, and so on.

Horizontally bedded rocks in an area of hills and valleys give patterns in which the rock boundaries are parallel with the contours. Because each boundary surface is a horizontal plane its outcrop at the surface will be a contour line itself, but not necessarily coinciding with the contour intervals of the map. Fig. 38*e*

Fig. 38. Geological map patterns. (*a*) Uniform dip of 30°. Width of outcrop is controlled by dip and bed thickness. (*b*) Symmetrical syncline. (*c*) Syncline plunging to the left. (*d*) Anticline plunging to the right. (*e*) Horizontal beds. Width of outcrop narrows on steeper slopes.

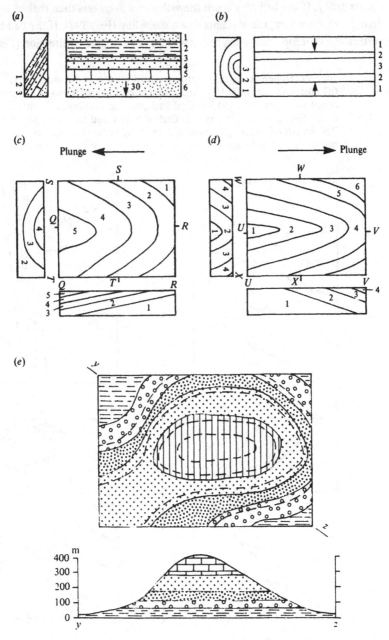

shows this structure, and the minor changes in boundary line direction when the ground surface is not flat are shown in Fig. 39a. These changes in direction when the boundaries cross a hill or valley are shown in magnified detail in Figs 39b, c, d. If the bed dips towards the head of a valley, the outcrop has a V-shape pointing up the valley. If the bed dips down the valley at a slope less than that of the gradient of the valley, the V points down the valley (Fig. 39d). If the bed is vertical the outcrop crosses the valley without any change of direction (Fig. 39b).

Fig. 39. Dependence of outcrop pattern on dip and topography. (a) Outcrop controlled by structure (plunging anticline) and topography (contours omitted). (b) Vertical bed crossing a valley. (c) Outcrop of a bed that dips up a valley. (d) Outcrop of a bed that dips down a valley. The width of outcrop is critically dependent on the angle of intersection between bed and valley slope.

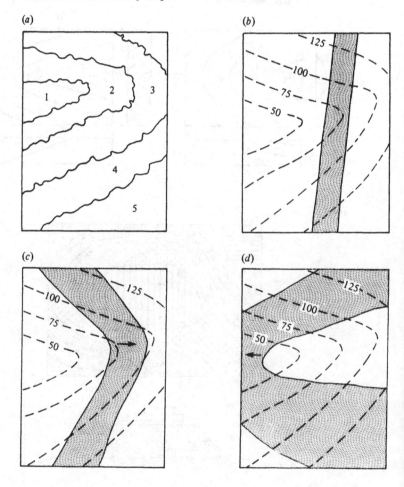

The above may be summarised by stating that if the beds are vertical the outcrops cross the hill and valley topography as straight lines, but if the beds are not vertical the outcrops are deflected as they cross valleys or hills, the more so as the dip angle decreases, until it becomes zero, when the beds run parallel with the contour lines on the map.

Short lengths of 15 cm wide planks of softwood such as deal are very useful for learning about the interpretation of fold structures. If they have been sawn through the centre of the tree trunk the grain will be in parallel, almost straight lines, but if they are not through the centre, i.e. flat sawn, they will have grain patterns similar to plunging anticlines and synclines. Look carefully at the end grain and that on the longer sides (Fig. 40). The effect is seen best when the wood is planed. Further similarities to geological maps and structures can be seen when the surface of the wood has been gouged with a wood carving tool to simulate valleys.

Faults cause lateral displacement of beds, shown in Fig. 41. A vertical fault plane outcrops as a straight line, like a geological boundary; otherwise non-vertical faults show changes in direction across the area, unless it is flat ground. Fig. 41*a* shows the effect of a fault which crosses the axis of a plunging anticline. The upthrow and downthrow sides of faults can be found on the map by applying the rule that younger beds are found on the downthrow relative to older beds on the upthrow side, as can be seen in Fig. 41*b*.

Faults often occur in sets which are more or less parallel to each other, and there may be two or more sets of fault directions in the area. It is easiest to find the relative direction of movement on faults by using the method of looking at the map in the direction down the dip. Turn the map so that the oldest beds are

Fig. 40. Dark summer wood layers in a block of softwood simulate fold structures in rocks.

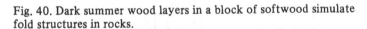
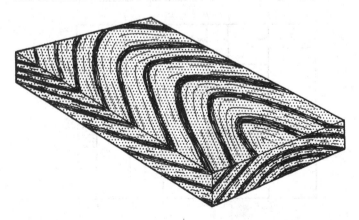

Fig. 41. Effect of faulting on outcrop pattern. (*a*) Normal vertical fault (*FF*) displaces outcrops of plunging anticline. Tick marks the downthrow side. (*b*) Vertical fault displaces outcrops of dipping beds. (*c*) Step faults in beds dipping away from viewpoint. Relative up and down movements are as seen from this side.

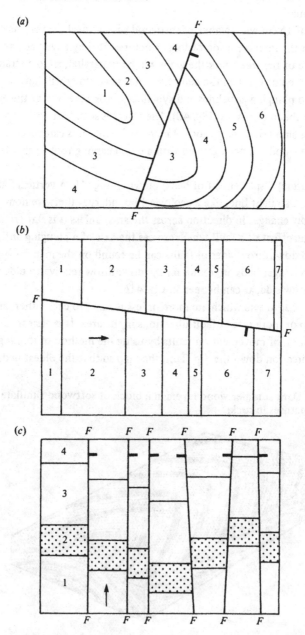

nearer the eye and the younger beds are further away, i.e. with the dip away from the viewpoint. Fig. 41c shows the application of this method to a series of faults, and if Fig. 38c is looked at again but turned sideways so that the apex of the U-shape is towards the eye, the downward structure of a syncline will be seen. If the line of sight makes an angle with the plane of the map equal to that of the dip of the beds the appearance of the structure will be approximately that seen in a vertical section through the rocks. Practice of this method, turning the geological map so that it is viewed down the dip in various places, will be a great help in identifying structures as anticlines or synclines, and seeing the relative displacements of faulted ground as if in a vertical section. The natural inclination is to look at the map so that the printed words are the right way up, but this does not help with the interpretation of the geological structure. After some practice, skill will be achieved and the structures will be seen in three dimensions immediately, without the need for turning the map. This method assumes, of course, that the relative ages of the rocks are known, and applies to structure-controlled outcrop patterns.

The shapes of igneous intrusions seen in plan view have been given in Figs 5 and 7. Note that igneous intrusion boundary lines intersect those of rocks older than the intrusions themselves, and that a series of intrusions of different ages can be inferred on the map by the way in which the boundaries intersect each other, as shown in Fig. 42.

The best way to understand geological structures is to practice drawing sections across geological maps and then to describe the history of the area in terms of periods of sedimentation, uplift, erosion, igneous intrusion and the sequence of faulting. Practice on the published geological maps or those in books

Fig. 42. Sequence of igneous intrusions. A sill (1) has been cut by two dykes (2), which in turn have been cut by a series of smaller dykes (3).

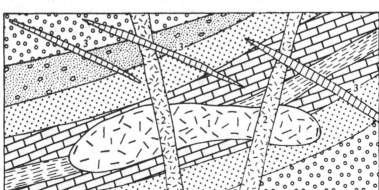

specially concerned with exercises on geological maps (problem maps) is very helpful.

Sections across geological problem maps are constructed in a sequence of steps, as follows:

1. Identify the rock types in the area by using the key. This may show the sequence of the rocks from oldest to youngest as in Fig. 43, or it may not. Look at the contours and visualise the topography of hills and valleys. The rock outcrop patterns are the result of intersection between a curved land surface and dipping or horizontal beds. Horizontal bedding is shown by geological boundaries parallel to the contours. The steeper the dip of the beds the less is the controlling effect of the topography on the outcrop pattern, the beds cross the map without much change of direction. The boundaries of vertical beds cross a map as parallel straight lines separated by distances equal to the thicknesses of individual beds (Fig. 39*b*). Curving patterns may also be caused by plunging anticlines and synclines (Figs 38*c*, *d*). Intersecting boundaries of sedimentary rocks indicate an unconformity. Faults cause displacement of outcrops on opposite sides of the fault (Fig. 41), or the disappearance of a bed on one side of the fault. Igneous intrusions intersect geological boundaries, unless they are sills occurring within single rock beds. This study of the rock types and outcrops should lead to the reconstruction of the geological history of the area.

2. Application of the above instructions to the problem map (Fig. 43) shows the following facts.

   There is an upper group or series of rocks (conglomerate and limestone) lying unconformably on a lower series (shale, lower sandstone, limestone, upper sandstone). There is high ground in the west of the map, and a valley running nearly diagonally through the centre of the map, and a spur of high ground in the north-east. The horizontal bedding of the upper series is shown by the way the geological boundaries run parallel to the contours; the unconformity is shown by the intersection of the bottom of the conglomerate bed with the boundary lines of the lower series, half way between the 700 m and 800 m contours.

   The way in which the boundary lines of the lower series intersect the contours shows that these beds are dipping. Any one of these boundary lines intersects contours of decreasing elevation in a south-easterly direction. This shows that the dip is to the south-east.

   A fault follows the line of the valley. It is a vertical fault because its outcrop is a straight line. The throw of the fault is found during the

Fig. 43. Geological problem map.

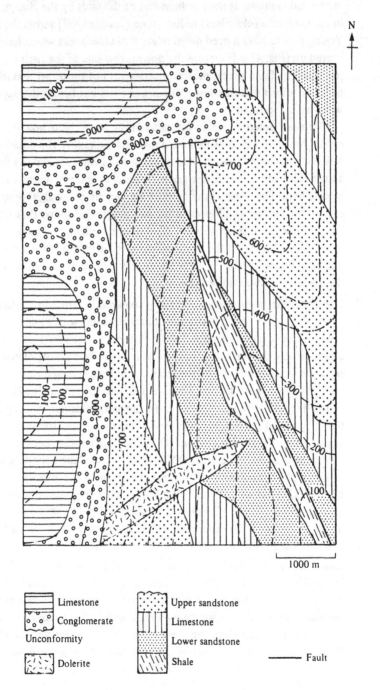

1000 m

Limestone

◦ ◦ ◦ Conglomerate

Unconformity

Dolerite

Upper sandstone

Limestone

Lower sandstone

Shale

Fault

accurate structure analysis described below, but the direction of the fault movement is shown at the northern end of the fault by the change from lower sandstone (older bed) to limestone (younger bed) across the fault. Younger rocks have moved down relative to older rocks which have moved up (Fig. 41). Therefore the downthrow side of the fault is on the east. The fault affects the lower series but not the upper. Faulting occurred after the formation of the lower series but before the upper series were deposited.

There is a dolerite igneous intrusion in the south. Note how it has caused the contours to be deflected towards the valley. The dolerite resists erosion more than the surrounding sedimentary rocks and forms a little ridge which crosses the generally smooth easterly slope of the ground here. The edges of the dolerite are two almost straight lines cutting the contours and intersecting at a point. This shows that the dolerite is in the form of a dyke, thinning and wedging out to zero to the north-east and that the intrusion occurred after the lower series had been formed, but before the upper series, because the boundary of the unconformity cuts the dyke.

3. The dip of the lower series is found by constructing stratum contours (also called strike lines). A stratum contour is a line which joins two points of equal elevation above survey datum on a defined plane in a bed, usually the top or the bottom of the bed. Two points of equal elevation and a third, not on the stratum contour joining the first two, are sufficient to give the dip. The stratum contour represents a horizontal line on the bedding plane, and a line perpendicular to this gives the direction of the dip. Fig. 45a shows this stage in the solution. At any point in the stratum contour AB the boundary plane between upper sandstone and limestone is at 600 m elevation. At C the boundary is at 500 m. Draw CD perpendicular to AB. The scale of the map gives $CD = 1000$ m. Between D and C the upper sandstone/limestone boundary has decreased 100 m in elevation – a fall of 100 m in 1000 m, or a dip of 1 in 10 ($5° 42'$). The direction of dip relative to north is found to be $130°$ by using a protractor.

4. Now mark on the map with distinct points all the intersections of bed boundaries and contours. Draw lines connecting these, as shown in Fig. 44. If the dip is constant and the beds have constant thickness in the map area, these lines will be parallel. Label each stratum contour with the appropriate elevation, for example U.S/L 500, using initials for simplicity. Some stratum contours can be given more than one label,

Fig. 44. Geological problem map. Stage 1, stratum contours drawn.

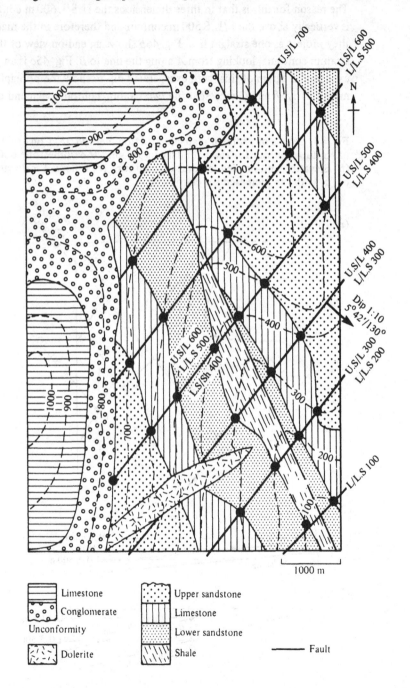

1000 m

Limestone

Conglomerate

Unconformity

Dolerite

Upper sandstone

Limestone

Lower sandstone

Shale

Fault

for example in Fig. 45*a*, *AB* is labelled U.S/L 600 m and L/L.S 500 m.
The reason for this is that in three dimensions the U.S/L 600 m contour
is vertically above the L/L.S 500 m contour and therefore in the map
they project as one straight line. Fig. 45*b* shows an end-on view of these
stratum contours, looking from *A* along the line to *B*. Fig. 45*c* is an
expanded diagram of this, with exaggerated dip to show the principle
more clearly. *P* and *Q* represent the two stratum contours seen end-on,

Fig. 45. Details of problem map construction method. (*a*) Construction
of stratum contour for the boundary upper sandstone/limestone 600 m.
(*b*) Elevation end-on view of stratum contours. (*c*) Enlarged scale view
of (*b*) with dip increased for clarity.

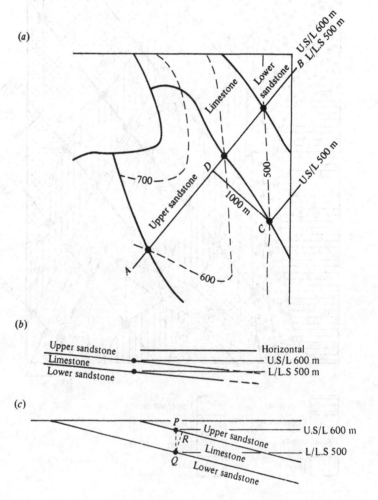

P vertically above *Q*. The difference in elevation of the two stratum contours (100 m) gives the approximate thickness of the bed between them. The true thickness is *QR*, but for dips of less than 1 in 5 it is generally accurate enough to take the difference in the vertical elevations. This gives the thickness of the limestone as 100 m. Similarly the stratum contour through the centre of the map, labelled U.S/L 600, L/L.S 500, L.S/Sh 400 m gives the thickness of the lower sandstone as 100 m.

It is essential to understand the method; it helps if the stratum contours are visualised as straight wires lying on the boundaries between beds. They are of course only construction lines and do not exist in reality.

Draw all the stratum contours as shown in Fig. 44, and label them accordingly. The thicknesses of the upper sandstone and the shale at the bottom of this series cannot be found by the reasoning just described, because there are no upper and lower boundaries for these beds. It is possible to make an estimate of thickness by considering the width of the outcrop of the 100 m thickness limestone just north of the dolerite dyke. It is about 600 m wide here, for a bed thickness of 100 m. The width of outcrop of the upper sandstone as far as the unconformity is about 1000 m, so the upper sandstone must be at least 120 m; otherwise there would be another boundary and another bed of different rock in this part of the map. Similar reasoning shows that the shale must be at least 70 m thick, measured towards the fault. At this stage note that the width of outcrop of a bed of constant thickness depends on the angle of intersection of the bed and the slope of the land surface. The smaller the angle of intersection the wider the width of outcrop, assuming constant bed thickness (Figs 38*e*, 39 and 46). One can now see the cause of the variation in the width of the lower sandstone along its outcrop on the south-west side of the fault. The bed has constant thickness and dip, but the direction of the slope of the ground changes from eastwards in the south of the map to south-eastwards (with the dip) at the north-west end of the fault. Changes of dip, or of bed thickness, will of course affect the width of the outcrop, but in simple geological problem maps these controlling factors are assumed to be constant.

The thickness of the conglomerate bed in the upper series can be found by interpolating between the contours. The base of this bed, the plane of the unconformity, outcrops halfway between the 700 and 800 m contours, that is at 750 m (assuming linear slope here). The 850 m level also marks the top of the conglomerate and base of the limestone, therefore the conglomerate is 100 m thick. The tops of the hills here are above 1000 m and therefore the limestone is estimated to be at least 150 m thick.

5. Now consider the fault. Its outcrop is a straight line, therefore the fault plane is vertical. The throw, the relative movement on either side of the fault, is found by reasoning as follows. The stratum contour on the south-west of the fault is labelled L/L.S 500 m for the lower sandstone/limestone boundary, but on the other side of the fault it is at 400 m. The reason is that all the beds on the north-east side of the fault have been downthrown 100 m. Check that this is so for the limestone/upper sandstone boundary.

This completes the analysis of the geological structure, which can be summarised as follows:

Upper series	Limestone	more than 150 m	horizontal beds
	Conglomerate	100 m	
Unconformity			
Lower series	Upper sandstone	more than 120 m	dip 1 in 10
	Limestone	100 m	(5° 42')
	Lower sandstone	100 m	N 130°
	Shale	more than 70 m	

Fault. Vertical, direction 335°, downthrowing NE 100 m.
Dolerite dyke. Maximum 500 m width, striking NE, post-lower series, pre-upper series.

6. Geological history. Formation of the lower sedimentary series in the sequence shale, lower sandstone, limestone, upper sandstone. Uplift of this series, tilting to the south-east, faulting, intrusion of the dolerite

Fig. 46. Relationship between width of outcrop of a bed (W), dip (θ), and slope of the ground surface, for constant bed thickness (T). (W') is the width of outcrop as projected on a map. The smaller the angle of intersection between ground surface and the bed the greater is the width of outcrop. (*a*) Vertical bed. (*b*) Dip into side of a hill. (*c*) Dip out from side of a hill.

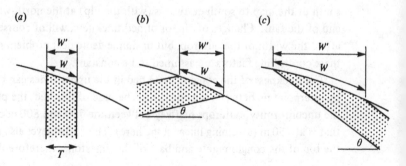

(a)   (b)   (c)

dyke. (There is no evidence to prove that this intrusion happened before or after the faulting.) Subsidence of the tilted sedimentary formation accompanied by erosion, then a further period of sedimentation when the conglomerate and the limestone were formed. A further stage of uplift, without tilting, formed land, then erosion followed, which has resulted in the present land surface.

7. The geology of the area is now illustrated by drawing a section, or more than one section if necessary, to show the rock formations in vertical perspective. Drawing the section requires a very sharp pencil, and the section is only inked when all the construction lines are finished and checked for accuracy. The direction of the section should be chosen so that it intersects all the outcrops of the various rock types. It may happen that a single straight line section will not do this, so that two or more sections are necessary. The section line *XY* is chosen because it intersects all the sedimentary rocks, the dolerite dyke, and the fault. This line is shown on the map (Fig. 47).

8. Draw a straight line *XY* of the same length in the middle of a sheet of paper, then choose and draw a suitable vertical scale. The vertical scale is often made three times the horizontal scale to make the topography and the beds appear more clearly, but this practice of course introduces an error in the dip, which is exaggerated by approximately three times. Dip can only be shown accurately when the vertical scale is the same as the horizontal, as in Fig. 48*a*.

9. On the map measure off the distances from *X* along the section line at which it intersects the contours. Transfer these distances to the section and measure upwards from the datum line (usually sea level). The distances can be quickly transferred by using the edge of a strip of paper and marking off the distances on it, and labelling the marks with the heights (Fig. 48*b*). Draw in the surface profile faintly, through the spot heights. The exact profile between the spot heights is not known; it can be convex, linear, or concave (Fig. 48*c*), and this will have an important effect on the accuracy of the section when the beds are drawn in, as described below.

10. Draw in the geological boundaries between the beds. Their positions are transferred from the map, as for the heights, by marking on the paper strip the points of intersection of boundaries with the section line. Mark the position of the fault at the same time. After a little practice all the information on heights, boundaries and faults can be marked on the paper strip at one time (Fig. 48*b*). It is best to indicate the rock types

Fig. 47. Geological problem map. Stage 2, line of section XY.

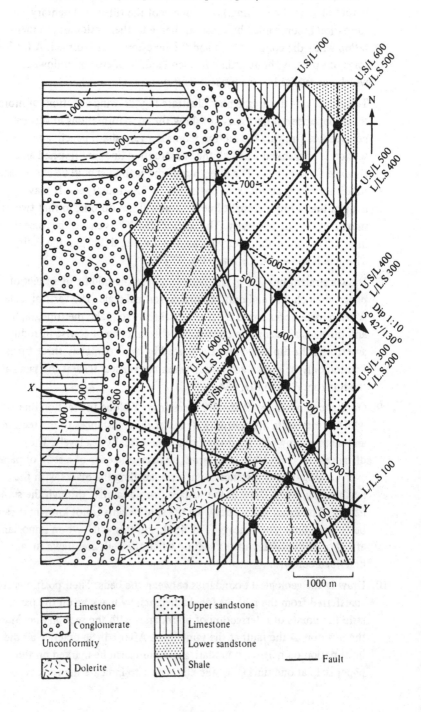

Limestone

Conglomerate

Unconformity

Dolerite

Upper sandstone

Limestone

Lower sandstone

Shale

Fault

Fig. 48. Geological problem map. Stage 3, drawing the section. (*a*) Construction of surface profile using spot heights. (*b*) Strip of paper used to locate spot heights, geological boundaries, stratum contours, and fault. (*c*) Cause of vertical error when locating geological boundary outcrops (magnified view). Three possible vertical positions of L/L.S boundary, all starting at the correct distance from section end. (*d*) Accurate location of geological boundary outcrop. (*e*) Completed section.

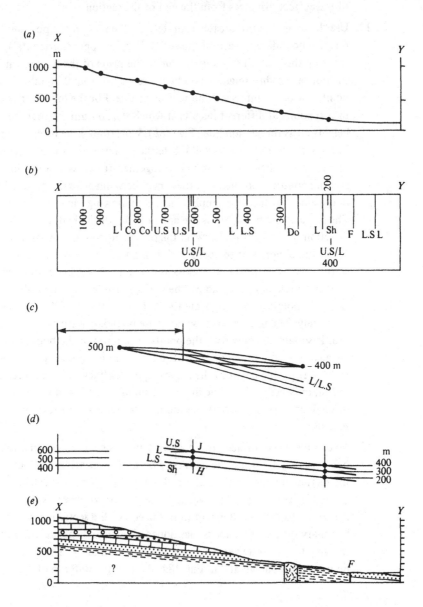

between the boundaries to avoid mistakes when completing the section. The positions of the outcrops of the boundaries are marked with a thin vertical line. Unless a boundary outcrop happens to coincide with a spot height, there is an uncertainty about the exact location of the point, caused by lack of information about the shape of the surface profile between the spot heights, as just noted. These vertical lines are really loci of the correct distances from the end of the section.

11. Use the paper strip to locate accurately on the section the position of one bed boundary in each of the series. For the upper (horizontal) series on this map this boundary can be the plane of the unconformity, the bottom of the conglomerate bed. Draw a horizontal line through the point at which it intersects the surface profile. For the lower series mark in two points at different heights at which the stratum contours for these heights cross the section line (Fig. 48b). When they have been transferred to the section these points will locate the geological boundary accurately, provided that they are not too close together. If necessary draw in more stratum contours on the map; these may be well above the ground level in the area, but they are simply a means of drawing the slope accurately. The slope of the boundary is the apparent dip of the beds along the direction of the section. Draw all the other bed boundaries, using the vertical thicknesses of the beds as shown on the stratum contours, starting with a vertical line of positioning dots (Fig. 48d). It does not matter if these points are above the surface profile; they are only construction points drawn to locate the vertical positions of the boundaries accurately. The intersections where these boundaries cross the surface profile should coincide with the points of outcrop of the boundaries already marked on the section. If the points do not coincide, the reason is that the profile between the nearest spot heights was not accurately drawn, only estimated while drawing a smooth curve, and the outcrop of the boundary may have been too high or too low, as shown in Fig. 48c.

12. To complete the surface profile accurately, using all the information, it is necessary to locate the boundary outcrop accurately. The point on the section at which a particular bed boundary outcrops is exactly located by two lines: (i) the vertical line through H, where XH is the measured distance on the map from the end of the section at X to the boundary outcrop (the locus mentioned above); and (ii) the other line showing the accurate vertical position of the boundary. The point of intersection is shown at J in Fig. 48d. Because the outcrop of the

boundary has now been fixed by these two construction lines, the
surface profile must pass through their intersection point (*J*), which
gives another spot height for surface profile. Locate all the other
boundary outcrops. One can then make a final improvement to the
surface profile, using the assumption that hard rocks (dolerite, sandstone, limestone on this map) tend to cause convex land surfaces, and
soft rocks (shale) cause concave surfaces.

13. For less accurate sketch sections it is sufficient to guess the surface
profile between the spot heights. Draw a smooth profile, then mark in
the boundary outcrop points on this profile, at the correct distances
from the end of the section. Then draw in the boundaries at the calculated dip angle for the direction of the section. There may be a vertical
error in the positions of the boundaries, but this only causes trouble
when the dip of the beds is nearly parallel to the slope of the ground;
outcrops may then be found to have been drawn above ground level
instead of on it, for the correct distances from the end of the section.
Some adjustment of height will remove this error.

The method described here can be used for solving problem maps in
general. The published books of problem maps contain a variety of maps
of increasing difficulty, including dip variation across the map caused by
folding of beds and reversal of dip across fold axes.

Geotechnical maps

These are also called engineering geology maps. Small-scale maps are
used for area surveys; plans made to 1:1250 and 1:500 are commonly used for
site surveys.

The mechanical properties of rocks and soils are more important to civil
engineers than the scientific descriptions and classifications based on chemical
and mineralogical principles. Geological maps give information about the type
of rock, but do not include information on the physical and mechanical aspects
like weathering, fissuring, although cleavage and other linear structures are
sometimes included for areas of highly-stressed and deformed rocks such as
schists. Since 1970 there has been an increasing demand for and the production
of special maps to show those properties of rocks and soils which are of special
interest to civil engineers.

One or more variables are shown on a map, or on a transparent overlay to be
placed on a topographical or ordinary geological map. A large number of variables
may be plotted and the student should read publications listed in the Bibliography.
The following list of properties of rocks and soils which may be plotted on maps

is not exhaustive, but will show the wide range of possibilities:

Hydrology: hydrogeological maps

Degree and depth of weathering

Depth to rockhead

Foundation requirements: stiff rafts, reinforced strips, normal strip foundations

Geological hazard: for example the probability of landslips or earthquakes

Slope instability, landslips in progress, abundance, susceptibility to sliding

Areas of high tectonic stress, caused by mountain-building movements

Discontinuities

Strength, deformability, porosity, permeability

Particle size distribution (soils)

Atterberg limits and plasticity index

Lithological descriptions of rocks, glacial and other forms of drift, and sometimes including thicknesses of deposits

Geomorphological maps: topography, and origin, age of the individual geomorphic units in greater detail than found in published topographical maps

Geodynamic factors: rates of erosion and sedimentation, and other processes such as landslides, seismicity, volume changes in soil (swelling coefficient), and sand dune movement

Fill (infilling): particularly in urban redevelopment areas. Towns grow upwards as well as outwards during their histories. Natural hollows in the land, disused quarries and clay pits are filled with waste and demolition rubble. The age of the fill should be noted where known; recently-filled areas are notorious for settlement, and special design is necessary, for example raft foundations, even for domestic dwellings

Documentation maps: record of sources of geotechnical information for the area, borehole distribution and type, dates, mineshafts, records of mineral extraction

Isopachyte maps: lines joining points of equal thickness of deposit or type of rock

Slope categories: angles of slope, 0-5°, more than 5° and less than 10°, and so on. Reverse slopes should also be recorded. Unit weight and natural moisture content can be recorded on slope maps because of their close relationship

Mining subsidence: for areas where there has been coal or metal mining over a long period. Ancient mining areas are full of small pits, trial holes backfilled or not, bell pits, shallow adits. There will also be waste tips, and some of these may cover old shafts so that no clue to their presence can be seen on the surface

Stereographic plotting (stereonet) of the dip of joint planes, faults, and other planar elements in the rock mass. Poles to planes are plotted, as angle of dip and azimuth.

Engineering geology zoning: stability maps are an example. They are prepared on small-scale maps for a particular type of engineering work, for example highways, tunnels, dams, new towns, to help in the choice of site or route, so that areas of expected rock trouble may be avoided if possible.

The very large number of variables which may be measured and plotted on maps and plans requires that a classification of geotechnical maps be made. The UNESCO publication quoted in the Bibliography (Anon., 1976*a*) includes such a classification.

One major difficulty in the preparation of geotechnical maps is the showing of variable thickness multilayer deposits on a variable type of bedrock having an undulating rockhead surface (Fig. 49).

Because of the large number of geotechnical properties which are measured during a survey there is a danger of including too many on a single map so that the user is confused by all the information which is given. There are many ways of recording the information on maps and a good account of the subject is given by Monkhouse and Wilkinson (1971), quoted in the Bibliography. Several systems

Fig. 49. Complex superficial deposit above an uneven rockhead surface.

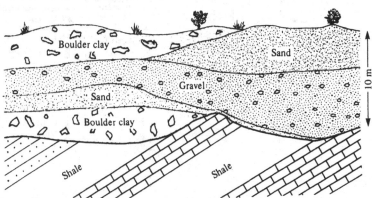

Fig. 50. Geotechnical map. (Based on 'The preparation of maps and plans in terms of engineering geology', Anon., 1972).

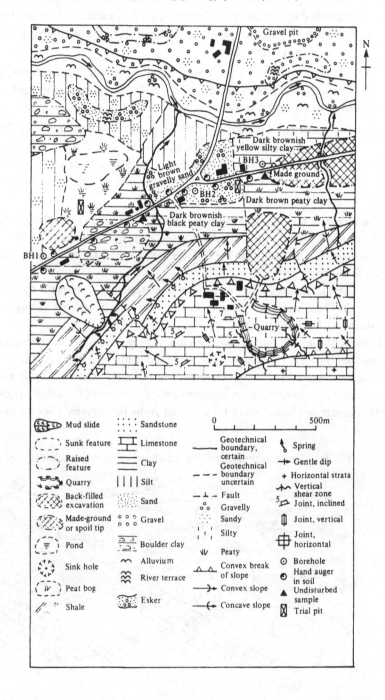

of symbols are in use, although there have been attempts at standardisation. The map user may have little experience of interpreting these maps, whereas the compiler (the site investigation engineer or geologist) may spend all his time on this type of survey and making maps from the results. Clarity of presentation of the information is essential.

The symbols used on geotechnical maps include colouring, shading, parallel lines, cross-hatching, letters and figures to give actual measured values. There are also special symbols to indicate positions of boreholes, cleavage direction, dip. A comprehensive list is given in 'The preparation of maps and plans in terms of engineering geology' (Anon., 1972), quoted in the Bibliography. Fig. 50 shows some of the symbols used in geotechnical maps. The student should examine this map critically and decide whether or not it shows too much information, and how it might be improved with respect to clarity.

6   Engineering description of rocks*

Rocks may be described for scientific purposes in terms of the mineral composition and mineral texture, chemistry, and origin. The classification of rocks in terms of origin leads to the well-known groups, sedimentary, igneous, and metamorphic rocks. This scientific classification is best suited for the purposes of geology, but has not proved satisfactory when the mechanical strength of rock is the main object of the classification and description. Geological descriptions do not stress the weathering state of rocks; it is in fact difficult to determine the mineralogy and chemistry of highly-weathered rocks, and geologists tend to concentrate on finding fresh specimens.

The primary requirement for engineering purposes is a report on the mechanical strength and behaviour of rock masses, and the true geological name of the rock is of secondary importance, but this is still retained in the engineering descriptions because the chemistry and mineralogical composition of a rock determine its weathering behaviour under different climatic conditions. The rock mass is described in terms of indices. Some of these are purely descriptive, for example colour, texture, weathered state; the others are measured by *in situ* or laboratory tests, for example porosity, strength, specific gravity, or are based on large scale structures within the rock mass, jointing, cleavage, planes of weakness, and bedding planes in sedimentary rocks. Rock mechanics is the science which attempts to describe and predict the properties and behaviour of rock masses. Soil mechanics deals with clay and granular material which is generally known to geologists as 'soft rock'.

For engineering purposes the description is divided into two parts: description of the rock material, and description of the rock mass.

Rock material description
Rock is a naturally-formed mixture of certain mineral species. In a hard rock these are firmly bonded together and the shapes, sizes, and orientation

* The information given in this chapter has been based on 'The description of rock masses for engineering purposes', *Quarterly Journal of Engineering Geology* 1977, Vol. 10, 355–388 (the Report of the Geological Society Engineering Group Working Party).

(aligned or random) determine the strength of the rock. Unconsolidated decomposed rock also consists of distinct minerals, mostly species of clay minerals, hydroxides of iron, and variable amounts of partly-decomposed original minerals. Some minerals are very resistant to decomposition in some climates, others decompose relatively quickly. The whole rock mass is a mixture of undecomposed original (primary) minerals and the secondary minerals produced by the chemical processes of decomposition or alteration.

The following parameters can be used for the description of rock material: colour, grain size, texture, specific gravity, hardness, weathering state, strength, primary permeability, seismic velocity, modulus of elasticity, swelling coefficient, slake durability, rock name.

The rock mass is described by noting the following structural properties: breaks (discontinuities), bedding planes in sedimentary rocks and lamination in others, strength, deformation modulus, secondary permeability (hydraulic conductivity), seismic velocity.

Rock material indices

Many of these indices are qualitative descriptions and it is necessary to define the properties in detail to obtain uniformity.

Descriptive indices
Rock type (petrological species)

The recommended rock types for description purposes are given in table 3. These rocks have been defined and described briefly in Chapter 2. The numbers in brackets after each rock type in the table are code numbers based on position in a series of columns (first digit in the number).

Colour

This consists of three components: hue, chroma, and value. Hue is the parameter which is generally known as colour (e.g. red), chroma is a qualifier (e.g. brownish) added to the hue, and value is a broad parameter (light or dark). Table 4 gives the terms used.

Grain size

This is the same as that used in the description of soils and other unconsolidated material. The method used for sizing is based on a series of standard sieves, but when solid rock is described it has to be cut into thin, transparent sections for examination in transmitted light under a petrological microscope. The measurement of grain size in thin sections is difficult because the section may pass through the corner of an individual crystal so that its full size does not appear in the section. Transverse sections of elongated crystals will not show their true length. Statistical methods are used to measure grain size in thin sections of rocks. Crystals of 60 μm size and above can be seen without the aid of micro-scopes. A rock that contains grains of less than 60 μm size is classified as fine-grained. Some rocks consist of minerals of two distinctly different sizes and are called porphyritic. Table 5 gives the standard grain sizes.

Texture, fabric, structure
   These terms are used in different ways in geology textbooks so it is necessary to define the meanings as used for engineering purposes. Texture refers to the individual mineral grains, their size, shape, and degree of crystallinity. Fabric

Table 4. *Rock colour. (Taken from 'The description of rock masses for engineering purposes' (Anon. 1977))*

Value	Chroma	Hue
light	pinkish	pink
dark	reddish	red
	yellowish	yellow
	brownish	brown
	olive	olive
	greenish	green
	bluish	blue
	greyish	white
		grey
		black

Table 5. *Grain size. (Taken from 'The description of rock masses for engineering purposes' (Anon., 1977))*

Term	Grain size	Equivalent soil grade
very coarse-grained	>60 mm	boulders and cobbles
coarse-grained	2–60 mm	gravel
medium-grained	60 μm–2 mm	sand
fine-grained	2–60 μm	silt
very fine-grained	<2 μm	clay

(grains >60 μm diameter can be seen with the unaided eye)

is the relation between the grains, the way in which they are arranged within the rock. In some rocks the minerals are in random orientation, but in many metamorphic rocks flat-shaped or elongated mineral crystals are arranged in parallel orientation (slate and schist). Structure refers to the larger scale features in rocks, for example some rocks have the same appearance over the whole mass (homogeneous), others have layers of different mineral composition and these give the rock a banded or striped appearance, which is called foliation.

Texture. Crystalline, crypto-crystalline (crystalline when seen under a microscope), granular (like sugar), amorphous (no particular shape), glassy.

Fabric. Random orientation of minerals, schistosity (parallel orientation). The orientation of mineral grains can be determined by identifying and measuring the positions of their crystallographic axes with a petrological microscope. The procedure takes up much time and is not often used in geotechnical tests. Mineral orientation is however a determining factor in rock strength. Randomly-oriented mineral grains produce isotropic rock strength; anisotropy is caused by alignment of flat or elongated minerals, or foliation. There are various degrees of mineral alignment ranging from the very obvious, which can be measured with a protractor and compass, to an almost hidden partial alignment which can only be detected with the aid of a petrological microscope.

Weathered state

This is the result of mechanical and chemical processes in the Earth's surface or close to it, when original (primary) minerals are decomposed and other (secondary) minerals are built up. Solution processes may remove material from the body of the rock, leaving it porous. Decomposition of rocks that contain iron produces red, yellow, or brown secondary minerals and the presence of these in a rock indicates its state of weathering. Weathered rocks have less mechanical strength than fresh rocks. The top 10 m of rock in the ground is normally in various states of weathering, decreasing in intensity downwards, generally, but not always. There may be pockets of highly-weathered rock surrounded by only slightly weathered rock. Kaolinized granite masses often show this irregular weathering structure. Table 6 defines the weathering classification scheme. Alteration is the word used to describe definite mineralogical changes caused by weathering, for example the conversion of feldspar to clay minerals, kaolinite, etc. Study of rocks in thin section under the microscope reveals the early stages of mineralogical decomposition that are not apparent in pieces of rock. The student should note the important relationship between mineral decomposition and rock strength.

Table 6. *Weathering Classification. (After British Standards Code of Practice for Site Investigation, BS 5930:1981)*

Weathering grades of rock material		Weathering grades of rock mass		
Term	Description	Term	Description	Grade
Fresh	No visible sign of weathering of the rock material	Fresh	No visible sign of rock material weathering; perhaps some slight discolouration on major discontinuity surfaces	I
Discoloured	The colour of the original fresh rock material is changed and is evidence of weathering. The degree of change from the original colour should be indicated. If the colour change is confined to particular mineral constituents this should be mentioned	Slightly weathered	Discolouration indicates weathering of rock material and discontinuity surfaces. All the rock material may be discoloured by weathering.	II
		Moderately weathered	Less than half of the rock material is decomposed or disintegrated to a soil. Fresh or discoloured rock is present either as a continuous framework or as corestones	III
Decomposed	The rock is weathered to the condition of a soil in which the original material fabric is still intact, but some or all of the mineral grains are decomposed	Highly weathered	More than half of the rock material is decomposed or disintegrated to a soil. Fresh or discoloured rock is present either as a discontinuous framework or as corestones	IV
Disintegrated	The rock is weathered to the condition of a soil in which the original material fabric is still intact. The rock is friable, but the mineral grains are not decomposed	Completely weathered	All rock material is decomposed and/ or disintegrated to soil. The original mass structure is still largely intact	V
The stage of weathering described above may be sub-divided using qualifying terms, for example 'partially discoloured', 'wholly discoloured', and 'slightly discoloured', as will aid the description of the material being examined		Residual soil	All rock material is converted to soil. The mass structure and material fabric are destroyed. There is a large change in volume, but the soil has not been significantly transported	VI

Strength

It is important to distinguish between the strength of a single piece of rock and that of the whole rock mass, which is largely determined by the frequency and orientation of discontinuities. *In situ* and laboratory tests are used to measure rock strength. The common tests are of unconfined compression strength, point load test, Schmidt rebound test, cone indenter test. Table 7 gives a classification of rock material strength. Note that mineral orientation causes the strength to vary with the direction of measurement, and the rock is then described as anisotropic. Rocks that have a well-defined grain (like wood) can be more easily split along the mineral grain than across it. This property is called cleavage when the oriented minerals are flat-shaped mica and chlorite crystals. The shear strength of rock samples can be found by machining samples to fit in the standard shear box used for testing soils. Shear strength along planes of discontinuities can also be measured.

Table 7. *Rock strength. (Taken from 'The description of rock masses for engineering purposes' (Anon., 1977))*

Term	Unconfined compressive strength $MN\,m^{-2}$ (MPa)	Field estimation of hardness
very strong	>100	very hard rock — more than one hammer blow required to break specimen
strong	50–100	hard rock — hand-held specimen can be broken with single hammer blow
moderately strong	12.5–50	soft rock — 5 mm holes made with sharp end of hammer
moderately weak	5.0–12.5	too hard to be cut by hand into a triaxial specimen
weak	1.25–5.0	very soft rock — material crumbles under hammer blows
very weak rock or hard soil	0.60–1.25	brittle or tough, broken in the hand with difficulty
very stiff	0.30–0.60*	soil can be indented by the finger nail
stiff	0.15–0.30	soil cannot be moulded in the fingers
firm	0.08–0.15	soil can be moulded only by strong pressure of fingers
soft	0.04–0.08	soil easily moulded with fingers
very soft	<0.04	soil exudes between fingers when squeezed in the hand

* The compressive strengths for soils given above are double the unconfined shear strength.

The hardness of rock mentioned in table 7 should not be confused with Mohs' scale of hardness (1-10) used in the description of minerals. Some rocks that have great surface hardness (e.g. obsidian or natural glass) resist indentation but may be very brittle and splinter when hit with a geological hammer because their impact strength is low.

Indices that can be determined by tests which require little or no sample preparation

Hardness

This is measured by the degree to which a steel hammer will rebound from a prepared rock surface. The Schmidt rebound hammer test gives a rebound number which can be correlated with the uniaxial compressive strength when the dry density is taken into account.

Durability

The slake durability test measures the resistance of rock to weakening and disintegration when immersed in water.

Porosity

This is defined as the percentage of void space in a rock: (volume of voids/total volume) × 100

$$n = \frac{V_v}{V_t} \times 100$$

Voids ratio

This is defined as the ratio (volume of voids/volume of solids) × 100

$$e = \frac{V_v}{V_s}$$

Density

The density of rock material is defined as the mass per unit volume, $g\,cm^{-3}$ or $Mg\,m^{-3}$. This is dependent on the specific gravity of the different mineral constituents of the rock or soil. Unconsolidated material, sand, silt, clay, consists mostly of quartz (S.G. 2.65), clay (approximately 2.70), mica (2.80-3.20) and iron oxides or hydroxides (3.60-4.00); and on the voids in the material, filled with air or water, or both. Consideration of these factors leads to four different specifications of density:

(1) The density of the solid mineral material, its mass per unit volume

(2) Dry density. The mass of the solids, with the voids filled with air only, per unit volume

(3) Saturated density. The mass of the material with the voids filled with water, per unit volume

(4) Bulk density. This is the mass per unit volume for the general case, when the voids are filled partly with air, partly with water. For this specification the moisture (water) content of the specimen is also measured and recorded.

Another test is used for gravel and sand, granular material which can be compacted to give a higher density than in the natural state. A minimum density value is found by pouring a known weight of material through a funnel into a measuring cylinder to find the volume it occupies when in its loosest state. The same sample is then poured into the cylinder in stages, each pouring followed by compaction with a hammer or rammer, in accordance with a specified procedure. This test gives the maximum density value.

The dry apparent specific gravity of rock is found by coating a weighed oven-dried specimen with paraffin wax and then immersing it in water and measuring the amount of water displaced, which gives the volume. The weight of the wax used, and its density, are taken into account when calculating the true volume of the sample.

Details of specific gravity and density tests have been described by Duncan (1969) and Ackroyd (1957).

Sonic velocity

This is a measurement of the velocity of sound in rock. It can be made in the laboratory by using an ultrasonic generator coupled to the specimen by transducers to transmit the sound into and out of the rock, which has been prepared by machining to a rectangular or cylindrical shape with flat ends for good sonic contact. The velocity of sound through rock is primarily dependent on Young's modulus of elasticity, and porosity. In general, the more solid the rock the greater the velocity of sound in it. In the rock mass however, discontinuities reduce the velocity considerably, both in specimens for laboratory testing and in rock in the ground. Field tests of seismic velocity in the rock mass are made by using an explosive charge or a dropped weight as the source of sound energy, and the usual geophysical technique using geophones and electronic signal recording.

To measure Young's modulus and Poisson's ratio the specimen must be specially prepared from cores by machining parallel flat ends. The specimen is hydraulically compressed in a triaxial compression machine. Measurements can be made on the specimen in an unconfined condition, surrounded by normal atmospheric pressure, or confined by surrounding the rock specimen with a steel cylinder connected to a supply of hydraulic pressure. The strain is recorded as the compressive stress is increased, and the process can be cycled.

Primary permeability

This is a measure of the amount and size of void space in the rock. Porous rocks can contain very small voids through which water cannot move because of the high surface tension effects in very narrow passages through the body of the rock. With larger voids the surface tension between water and mineral grains is less and the rock is more permeable. There is no direct connection between porosity and permeability. Secondary permeability refers to the whole rock mass, and is controlled by the density of discontinuities.

Rock mass description

The rock mass over the whole construction site may consist of one or more types of rock. These and the geological structures (joints, bedding, etc.) are measured and described by a standard procedure. The measurements are reported as rock mass indices.

Rock mass indices

Discontinuities are recorded as follows:

Type

Number of discontinuity orientations

Location and orientation

Frequency of spacing between discontinuities

Aperture (separation of discontinuity surfaces)

Persistence and extent

Infilling

Nature of surfaces.

A discontinuity is a plane of weakness within the rock mass, across which the rock has a low tensile strength, or along which it has a low shear strength. Some of these planes may be open, as in a joint, but they may be closed (no distinct break seen until the rock has failed under test). The planes include joints, cleavage, schistosity, foliation, veins, bedding planes, faults or open fissures caused by solution of soluble rock-forming minerals (e.g. calcite). Fig. 51 shows some types of discontinuity. The three-dimensional attitudes of these discontinuities are measured by recording dip angle and direction with a clinometer

Fig. 51. Discontinuities in rocks: (*a*) bedding planes and joints; (*b*) fault zone filled with fault breccia and clay; (*c*) quartz veins as isolated discontinuities in sandstone; (*d*) interconnected discontinuities; (*e*) detail of wavy discontinuity in rock bed; (*f*) roughness — stepped profile; (*g*) angular roughness on discontinuity caused by mineral grains; (*h*) steatite layer formed from olivine by shearing.

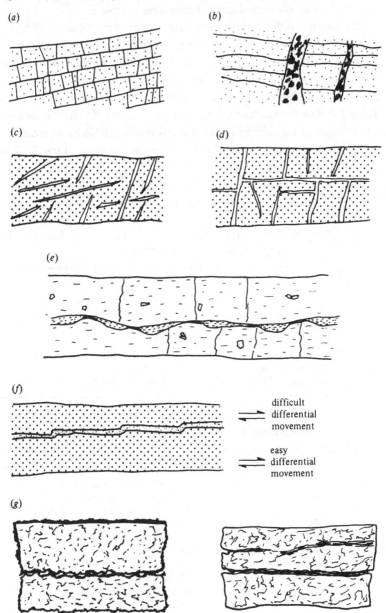

and a compass and the results are plotted on a stereographic projection. This will reveal any well-defined directions of structural weakness.

An index of the frequency of discontinuities is obtained by counting the number which intersect a line of suitable length to give a measure of the mean spacing. Measurements are made along three mutually-perpendicular axes. Discontinuity spacing can be described according to table 8.

The amount of discontinuity aperture is an important control of rock mass stability. Joints may be closed up (tight), or they may be open. The space between adjacent blocks may or may not be filled with decomposed rock (infilling), which is often clay, or a mixture of clay and rock pieces, or a mineral vein. The infilling should be described on a basis of the description of rock and soil given above. Table 9 defines the aperture of discontinuity surfaces. The unconfined compressive strength of the infilling can be assessed visually (table 7), or measured with a pocket penetrometer or vane tester for soils, or a point load test for harder material.

The total area of an individual discontinuity is known as its persistence, and is an important factor in the strength of a rock mass. It is difficult to measure;

Table 8. *Discontinuity spacing. (Taken from British Standards Code of Practice for Site Investigation, BS 5930:1981)*

Term	Spacing
very widely spaced	greater than 2 m
widely spaced	600 mm–2 m
medium spaced	200–600 mm
closely spaced	60–200 mm
very closely spaced	20–60 mm
extremely closely spaced	less than 20 mm

Table 9. *Aperture of discontinuity surfaces. (Taken from 'The description of rock masses for engineering purposes' (Anon., 1977))*

Term	Aperture (discontinuities) Thickness (veins, faults)
wide	more than 200 mm
moderately wide	60–200 mm
moderately narrow	20–60 mm
narrow	6–20 mm
very narrow	2–6 mm
extremely narrow	more than 0–2 mm
tight	zero

some major discontinuities, faults and major (master) joints, may persist through the whole length of the site, or cross part of it. The persistence of these discontinuity planes can be recorded by measuring their lengths in different directions along their planes. The discontinuities may vanish into solid rock (Fig. 51c), or terminate against other discontinuities (Fig. 51d). The surface of a discontinuity is important and may be described in terms of waviness, roughness and condition of the walls. They all have an effect on the shear strength along the discontinuity plane, together with the amount of water on the plane or in the infilling.

Waviness is defined as first-order roughness on the plane, waves on the rock surface that would not be broken off during movement along the plane. Waviness is measured in terms of amplitude and wavelength of the waves, using a tape (Fig. 51e).

Roughness is a second-order phenomenon caused by the inherent rock texture, grain size of minerals, decomposition of some minerals but not others. When there has been continual movement along discontinuity planes (e.g. faults) over millions of years at high temperatures (more than 200°C) and at the pressure existing at several kilometres below the surface, new minerals may form along the plane. These are flat-shaped (often steatite or clay minerals) and oriented parallel to the plane of the discontinuity. When the rock is broken these planes appear smooth and are said to be slickensided or polished. Table 10 defines the degrees of roughness. The surface irregularities sometimes show a distinct lineation and the coefficient varies with direction, for example when the surfaces consist of interlocking steps (Fig. 51f). This effect causes the movement in one direction to be easier than in the opposite direction. When the roughness varies with direction, this direction should be measured and reported.

The water content within discontinuities or infilling is very important in determining the stability of the rock mass.

Shape of naturally-formed blocks: when the three-dimensional aspects of a rock mass are considered, the rock can often be seen to have a tendency to

Table 10. *Roughness categories. (Taken from 'The description of rock masses for engineering purposes' (Anon., 1977))*

Category	Degree of roughness
1	polished
2	slickensided
3	smooth
4	rough
5	defined ridges
6	small steps
7	very rough

break up into blocks of regular or irregular shapes, depending on the pattern of the intersecting discontinuity planes. The blocks should be measured for size and shape as follows: the size classification is given in table 11; the orientations of long or short dimensions should also be recorded. The shape classifications are:

blocky   approximately equidimensional

tabular   one dimension much shorter than the other two

columnar   one dimension much larger than the other two

These shapes are shown in Fig. 52.

All these measurements and descriptions for the rock mass are recorded on specially-printed sheets designed for data processing. The descriptions and values can be converted to index numbers to simplify data storage.

The British Standards publication *Code of practice for site investigation* (BS 5930:1981) has replaced the earlier edition CP 2001 and includes all the standard tests made to determine the mechanical properties of hard and soft rocks.

Table 11. *Block size. (Taken from 'The description of rock masses for engineering purposes' (Anon., 1977))*

Term	Block size	Equivalent discontinuity spacings in blocky rock
very large	more than 8 m^3	extremely wide
large	0.2–8 m^3	very wide
medium	0.008–0.2 m^3	wide
small	0.002–0.008 m^3	moderately wide
very small	less than 0.002 m^3	less than moderately wide

Fig. 52. Shapes of naturally-formed pieces of rock after stress–relief expansion.

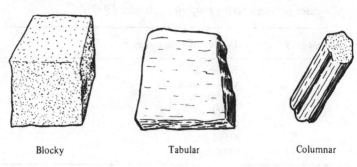

Blocky   Tabular   Columnar

Bibliography

Ackroyd, T. N. W. (1957). *Laboratory testing in soil engineering.* 233 pp. London: Soil Mechanics Ltd.

Anon. (1972). The preparation of maps and plans in terms of engineering geology. *Quarterly Journal of Engineering Geology,* 5, 293–381.

Anon. (1976a). *Engineering geological maps. A guide to their preparation.* 79 pp. Paris: The UNESCO Press.

Anon. (1976b). *Manual of applied geology for civil engineers.* 378 pp. London: Institution of Civil Engineers.

Anon. (1977). The description of rock masses for engineering purposes. *Quarterly Journal of Engineering Geology,* 10, 355–388.

Attewell, P. B. and Farmer, I. W. (1976). *Principles of engineering geology.* 1045 pp. London: Chapman and Hall.

Bell, F. G. (1978) (Ed.). *Foundation engineering in difficult ground.* 604 pp. London: Butterworth.

Bell, F. G. (1980). *Engineering geology and geotechnics.* 497 pp. London: Butterworth.

Bennison, G. M. (1975). *An introduction to geological maps and structures.* 64 pp. London: Edward Arnold.

Blyth, F. G. H. and de Freitas, M. H. (1974). *A geology for engineers.* (6th edn.) 557 pp. London: Edward Arnold.

British Standards (1981). *Code of practice for site investigation* (formerly CP 2001) BS 5930:1981.

Capper, P. L. and Cassie, W. F. (1966). *The mechanics of engineering soils.* 298 pp. London: Spon.

Carson, M. A. (1971). *The mechanics of erosion.* 174 pp. London: Pion.

Coates, D. R. (1976) (Ed.). *Geomorphology and engineering.* 360 pp. Stroudsburg, Pa.: Dowden, Hutchinson and Ross.

Duncan, N. (1969). *Engineering geology and rock mechanics.* 2 Vols. London: Leonard Hill.

Fookes, P. G. (1969). Geotechnical mapping of soils and sedimentary rocks for engineering purposes, with examples of practice from the Mangla Dam Project. *Géotechnique,* 19 (1), 52–74.

Hails, J. R. (1977) (Ed.). *Applied geomorphology.* 418 pp. Amsterdam: Elsevier Scientific Publishing.

Hoek, E. (1973). Methods for rapid assessment of the stability of three dimensional rock slopes. *Quarterly Journal of Engineering Geology,* 6, 243–255.

Hoek, E. and Bray, J. W. (1974). *Rock Slope Engineering.* 231 pp. London: The Institution of Mining and Metallurgy.

Jaeger, J. C. and Cook, N. G. W. (1976). *Fundamentals of rock mechanics.* (2nd edn.) 585 pp. London: Chapman and Hall.

Monkhouse, F. J. and Wilkinson, H. R. (1971). *Maps and diagrams.* 522 pp. London: Methuen.

Moseley, F. (1979). *Advanced geological map interpretation.* 80 pp. London: Edward Arnold.

Ollier, C. D. (1969). *Weathering.* 304 pp. Edinburgh: Oliver and Boyd.

Read, H. H. and Watson, J. (1968). *Introduction to geology. Principles. Vol. 1.* (2nd edn.) London: Macmillan.

Roberts, A. (1977). *Geotechnology. An introductory text for students and engineers.* 364 pp. Oxford: Pergamon Press.

Simpson, B. (1968). *Geological maps.* 112 pp. Oxford: Pergamon Press.

Way, D. S. (1973). *Terrain analysis. A guide to site selection using aerial photographic interpretation.* 392 pp. Stroudsburg, Pa. Dowden, Hutchinson and Ross Inc.

Zaruba, Q. and Mencl, V. (1976). *Engineering geology. (Developments in geotechnical engineering. Vol. 10.)* Amsterdam: Elsevier.

Index

acid igneous rocks 19
acid test 40
agglomerate 25, 29, 30
albite 12
alkali-aggregate reaction 10
alluvial deposits 6
alluvium 48
amphibole 12
andesite 26
angle of friction 81
anhydrite 14, 31
anisotropic strength 73
anorthite 12
anthracite 33
anticline 60
apparent dip 62
arenite 35
asbestos 12
augite 11

basal conglomerate 72
basalt 26, 73
basic igneous rocks 19
batholith 22, 72
bauxite 41
bed 60
bedrock 6
biotite 12
blocks, naturally formed 129, 130
blown sand 54
bole 29
boss 72
boulder clay 52
braided rivers 50
breccia 34
bulk density 125
buried karst 42
buried river channels 51

Cainozoic era 8
calcarenite 40
calcareous rocks 40
calcite 13
    in joints 70

calcirudite 40
calcisiltite 40
carbonate minerals 13
caves in limestone 41
chalk 33, 54
chemical composition of Earth's crust 14
chemical factors in rock decay 5
chert 31
chlorite 13, 44
clay 39, 80, 81
clay minerals 13
    swelling 13
claystone 37
clay-with-flints 54
cleavage in slate 43, 45
clinometer 60
coal 33
coastal deposits 57
coefficient of friction 81
cohesive soil 81
colour of rocks 119, 120
columnar structure 27
conglomerate 34
continental drift 2
cotangent method of dip calculation 66
country rock 22
cupola 72
current bedding 36

decomposition of rocks 5, 121
deformation of rock 66
density of rock 124
detritus 15, 31
diabase 26
dilatational jointing 73
diopside 11
diorite 25
dip 60, 91
discontinuities 1, 4, 27, 42, 65, 119, 126–30
    aperture 126, 128
    persistence 126, 128, 129
    roughness 127, 129
    spacing 126, 128
    waviness 129

dolerite 26
dolomite 13
dolomitic limestone 40
dome 60
drainage basin 50
drift 47
'drift' geological maps 90, 95
drowned river valleys 48
drumlin 52
dry apparent specific gravity 125
dry density 125
dune bedding 36
duricrust 47
dyke 27, 72

Earth
    age of 1
    history 7
    internal movements 8
    internal structure 1, 4
earthquakes 1
engineering geology maps 113
enstatite 11
eras, geological 7
erosion, 17
    cycle 75
esker 52

fabric of rocks 121
faults 3, 68–70
    gouge 70
    heave 68
    plane 68
    shift 68
    throw 68
    zone   70
feature mapping 91
feldspar 12
feldspathoid minerals 12
felsite 22
ferromagnesian minerals 26
fill 57
flank of fold 66
flint 31
flocculated clay 39
floodplain 50
folds 65–8
    axes 66
    flanks 65
    limbs 65
foliation 18, 42, 44, 121
formation, rock 60
fossils in limestone 40
freestone 41

gabbro 26
garnet in schist 44

geological boundary 91, 96
geological column 7, 8
geological maps 90–113
geomorphology 74
geotechnical maps 113–117
glacial deposits (glacial drift) 6, 51
glauconite 13
gneiss 44
grain of rock 62
grain size in rocks and soils 18, 119, 120
granite 19
granular soil 81
greywacke 36
grit 35
gypsum 13, 31

halite (rock salt) 31
halloysite 13
hardness of rock 124
head 52
heave of a fault 68
history of the Earth 6
hornblende 12
hornblende–schist 44
hornfels 22
hydrothermal mineral veins 14

ice, as agent of erosion 5
igneous rocks 15, 19–31
illite 13
in situ tests 4
indurated sediment 36
inselberg 24
instability in rock 74, 77
intermediate igneous rocks 19
internal structure of the Earth 1
iron hydroxides 5
iron minerals 14
isoclinal folding 65
isotropic strength 73

joints 4, 70, 73

kame 52
kaolinite 13
kaolinization 24
karst 42
    buried 42
knickpoint 77

lamination 36
landforms 73
    history 75
landscape 73
laterite 24, 54
lava, 27
    flow 15

leucite 12
lignite 33
limb of fold 65
limestone 40
limonite 14
loess 56
longshore drift 57
lutite 40

made-ground 46, 57
magma 15, 73
magnetite 14
mantle 1
maps
    geological 90–113
    geotechnical 113–117
marl 39
marram grass 54
meanders 50
Mesozoic era 8
metal-bearing minerals 14
metamorphic aureole 22
metamorphic rocks 42–46
metamorphism
    contact 22
    dynamic 16
    dynamothermal 16
    thermal 16, 22
mica minerals 12
microfossils 40
microgranite 22
migmatite 22, 44
minerals
    decomposition 10
    definition 10
    rock-forming 10, 11–14
    specific gravity 10
montmorillonite 13
moraine 52
mountain-forming processes 3
mudflow 57, 78
mudstone 36
muscovite 12
mylonite 46

nepheline 12
normal fault 68

obsidian 26
ocean-floor spreading 3
olivine 11
oolite (oolitic limestone) 41
organic sedimentary rocks 31
orthoclase 12
orthopyroxene 11
outcrop 4, 60, 90
    patterns on map 96

overburden 6
ox-bow 50

Palaeozoic era 8
patina 33
peat 57
pegmatite 22
peneplain 75
peridotite 30
periods, geological 7
permeability
    primary 119, 126
    secondary 119, 126
phyllite 45
pitchstone 26
plagioclase 12
planar slide 84
plateau gravel 56
Pleistocene ice age 77
plunging fold 96, 99
pluton 72
porosity 124
porphyritic fabric 18, 22
pot-holes 41
Pre-Cambrian era 8
primary period 8
problem maps 102–113
pyrite 14
pyroclastic rocks 25, 30
pyroxene 11
pyroxenite 30

quartz 10
    in joints 70
    veins 13
quartzite 13, 36
Quaternary period 8

Recent period 9
rejuvenation of landforms 76
residual deposits 54
reverse fault 68
rhyolite 22, 25
river terraces 48
rock bolting 84
rock falls 81–5
rock series 60
rock types (table) 20, 21
rockhead 6, 48, 54
rocks
    igneous 15, 19–31
    metamorphic 16, 42–6
    sedimentary 15, 31–42
rotational slide 86
roughness 127, 129
rudite 34

saline deposits 31
sand bars 57
sand dunes 54
sandstone 34
saturated density 125
schist 44
sea-level changes 76–7
secondary period 8
sedimentary rocks 15, 31–42
seismic areas 3
series, rock 60
serpentine 30
shale 37
shift of a fault 68
siderite 13
silica 5, 10, 12
silicate minerals 11
siliceous rocks 12
sill 27, 73
silt 39
siltstone 36
sinter 47
slate 45
slickensiding 68
slip on fault plane 68
slope
　angles 74
　movement 74, 77
　of ground 74
　related to rock type 75
smectite 13
soil 1
　creep 57
　formation 17
'solid' geological maps 90. 95
solifluction deposits 57
solution channels in limestone 41
solution pipes 54
sonic velocity in rock 125
specific gravity
　Earth 1
　rocks 16
spits 57
spring sapping 80
stage, geological time 8
stratum 60
　contour 104
strength of rock 123

strike 61
　line 104
structure, geological 60
　igneous rocks 72
summit accordance 77
superficial deposits 46
superficial zone 6
swelling clays 13
syenite 19
syncline 60

tachylite 30
talc 30
tear fault 68
terraces, river 48
Tertiary period 8
texture of rocks 121
throw of a fault 68
thrust fault 68
toppling 84
tor 24
tourmaline 19
transcurrent fault 68
translational slide 84
triaxial test 125
tuff 22, 25, 29, 30

ultrabasic rocks 19, 30
unconformity 70–2

valleys 48–51
varves 52
vermiculite 13
vesicles 29
vesicular texture 29
voids ratio 124
volcanic ash 29
volcanic dust 15
volcanic mudflow 30
volcanoes 15

waviness in discontinuities 129
weathered state of rock 121
weathered zone 5, 6
weathering 5, 17
　classification 122
wind, as agent of erosion 5
wrench fault 68